Hands-on Science

LIFE SCIENCE

Stephen A. Martin

and   Joseph Starowicz

The Peoples Publishing Group, Inc.

Cover Credits

Toadstool . . . . Sarah Sheffield White
Blackbar Soldierfish. . . . . Al Grotell
Human Chromosomes . . . . Phototake/Dr. Ram Varma
Hydra . . . . Panographics/L.S. Stepanowicz
White-footed Mouse . . . . Dwight R. Kuhn
Eastern Tailed Blue (Butterfly) . . . . Richard S. Todd
Water Lily . . . . Robert Maust
Frog . . . . Allen Ruid
Earth . . . . Index Stock International

ISBN 0-88336-854-4

© 1988

The Peoples Publishing Group, Inc.
230 West Passaic Street
Maywood, New Jersey 07607

Printed in the United States of America.

14 15 16 17 18 19 20

Table of Contents

Life Science

The planet Earth is rich with life and mystery. You are sharing the planet with more than one and a half million other forms of life. Some are plants, some are animals, and some are neither plants nor animals. This book will take you on a tour of these other forms of life. It will take you on a tour of your own body and how it works. You'll also find out how dependent all these life forms are on each other and on the environment. You'll discover that even the tiniest form of life has a big role to play on this small planet we call home.

6

Unit 1.
Biology and its tools

Welcome to the world of science. In this unit, you will begin to answer questions from a scientist's point of view. How does a scientist define life? What is a living thing? What is a non-living thing? How do they differ?

You will also begin to use the scientist's tools. You will learn how to use a microscope. You will see things that are too small to see with your naked eye.

Gail Denham

Jay Davis

Carl Purcell

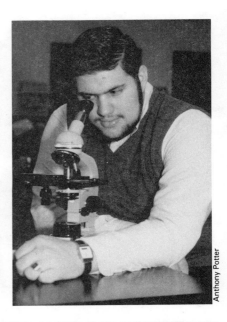

Anthony Potter

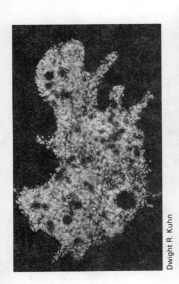

Dwight R. Kuhn

Jay Davis

1. What is life?

Biology is the scientific study of living things. **Biologists** are scientists who study living things. Suppose you are a biologist looking at the pictures on this page. You would say that the plant and deer have life. You would say that the chair does not have life.

But what is life? What is a living thing? What is a non-living thing? Biologists have tried to answer those questions. They say that living things have certain qualities in common. List the qualities you think living things have in common.

_____

_____

_____

_____

_____

_____

Non-living and living things

E. Cole

Anthony Potter

Anthony Potter

Now look at the biologists' list of the qualities that all **organisms** share. (Organisms are simply living things.) See how closely your list matches theirs.

1. Living things need energy. They get energy from the foods they eat. (Energy is the ability to move or do something.)
2. Living things are made up of **cells**. A cell is the smallest living part of a plant or animal. It is the basic unit of life. (Some living things are made of only one cell. Other living things are made of billions of cells.)
3. Living things grow.
4. Living things have a life span. They are born, live their lives, then die.
5. Living things are able to reproduce things like themselves.
6. Living things respond or react to conditions outside of themselves. These conditions can be heat, light, cold or touch.

These elk have all of the qualities of living things listed above.

Chapter checkup

Write **T** for true or **F** for false in the blanks below.

_____ 1. Scientists who study life are biologists.

_____ 2. Biology is the scientific study of the weather.

_____ 3. The cell is the basic unit of life.

_____ 4. Living things do not need energy.

_____ 5. Living things grow.

Science words

magnifying glass
lens
microscope
simple microscope
compound microscope

2. The microscope: the biologist's tool

The famous detective Sherlock Holmes solved crimes by finding clues. Often he would use a magnifying glass to find these clues. A **magnifying glass** makes an object look larger.

The lens

The most important part of a magnifying glass is its lens. A **lens** is a curved piece of glass. It makes objects look larger than they really are. Look at the picture below. Notice the size of the object. Notice how much larger the lens makes the object appear. When you look through a magnifying glass, you don't see the object itself. You see a larger image of it.

Anthony Potter

Anthony Potter

A magnifying glass makes objects appear several times larger.

Try this #1

Material: clear plastic wrap
1. Stretch a piece of clear plastic between your hands.
2. Have another student or your teacher put one drop of water in the center of the wrap.
3. Look through the drop of water at the letters in this book.
4. Describe how the letters looked through the drop of water.

_____

You have just created a simple lens. The letters in your book should appear larger through the water on the plastic wrap.

Microscope

A **microscope** is a device that makes objects appear larger. The word *micro* means "very small." The word *scope* means "to look at."

A **simple microscope** has just one lens. A magnifying glass is a simple microscope. For some people, eyeglasses are a simple microscope. The simple microscope is a very useful tool, but there are things so small even the best simple microscope is not powerful enough.

To see very small objects you need a **compound microscope**. A compound microscope has two or more lenses. The microscope you will use in class is a compound microscope.

Look at the drawing of a compound microscope. Before you read any further, think about each part of the microscope and its uses. Then read on.

An early microscope

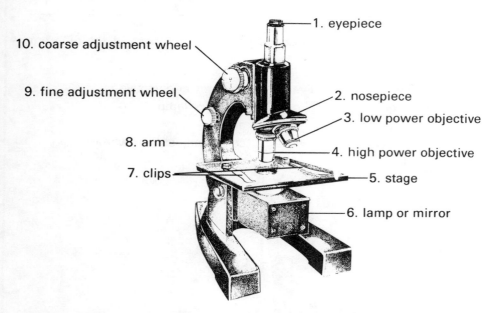

10. coarse adjustment wheel
9. fine adjustment wheel
8. arm
7. clips
1. eyepiece
2. nosepiece
3. low power objective
4. high power objective
5. stage
6. lamp or mirror

Parts of the compound microscope and their uses

1. The **eyepiece** is the part you look into. It holds the top lens of the microscope.
2. The **nosepiece** holds the bottom lenses of the microscope. It allows the lenses to turn so that either lens can be used.
3. The **low power objective** is one of the bottom lenses of the microscope. Together with the lens in the eyepiece it magnifies the object.
4. The **high power objective** is the other bottom lens of the microscope. It also is used with the lens in the eyepiece to magnify the object. This lens is more powerful than the low power objective.
5. The **stage** is the platform that holds the object you are looking at.
6. A **lamp** or **mirror** throws light onto the object you are looking at.
7. **Clips** hold the object you are looking at steady.
8. The **arm** connects the upper and lower parts of the microscope. It can be used as a handle when you carry the microscope.
9. The **fine adjustment wheel** is used to sharpen the image.
10. The **coarse adjustment wheel** is used to bring the object roughly into focus.

Try to become familiar with the parts of the microscope. Learn what these parts do. In the next chapter you will actually use a microscope.

An electron microscope is a very powerful microscope used by scientists.

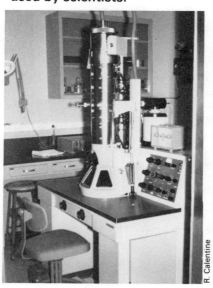

Chapter checkup

A. Fill in the blanks.

1. Two types of microscope are the _____ and the _____.

2. A piece of curved glass is called a _____.

3. An example of a simple microscope that is usually held in your hand is a
   _____.

4. The word *micro* means _____.

5. The word *scope* means _____.

B. Write the names of the microscope parts in the blanks in the diagram below.

low power objective	high power objective
stage	mirror or lamp
eyepiece	arm
fine adjustment wheel	coarse adjustment wheel
clips	nosepiece

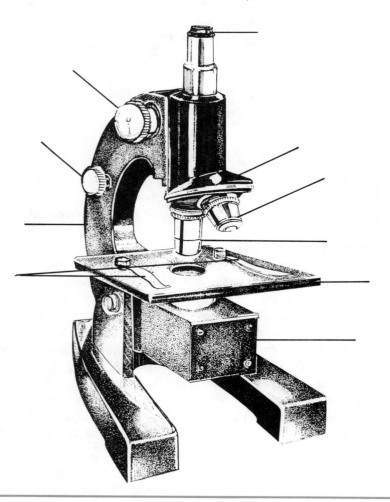

3. Using your microscope

Anthony Potter

Most compound microscopes have three lenses. The first lens is in the eyepiece. On the eyepiece you will find a number and the letter X. On many microscopes you will see 5X on the eyepiece. This means that the lens in the eyepiece makes things look five times larger than they actually are.

The other two lenses are in the objectives. The number on the low power objective will always be smaller than the number on the high power objective. On many microscopes the low power objective will have 10X on it. The high power objective will have 43X on it.

In a compound microscope, the eyepiece and one of the objectives are always used together. To find the magnifying power, simply multiply the power of the eyepiece by the power of the objective. For example, suppose the power of the eyepiece was 5X and the power of the objective was 10X. Multiply 5 times 10. The magnifying power is 50X. So when you are looking through the microscope at that setting, objects will look 50 times larger than they really are.

Human hair
magnified 100 times

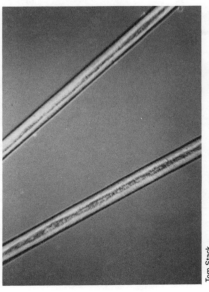

Tom Stack

power of the eyepiece is 5X		power of the objective is 10X		object is magnified 50 times
5	x	10	=	50

13

Things to do

Find the low and high power of this compound microscope.

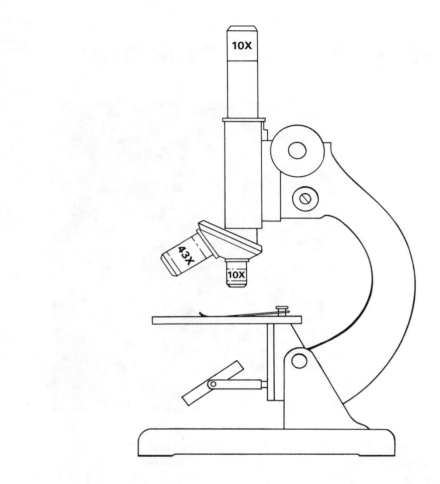

1. _____ X _____ = _____ times.
      (power of   (power of the   (object is
      the eyepiece)   low objective)   magnified)

2. _____ X _____ = _____ times.
      (power of   (power of the   (object is
      the eyepiece)   high objective)   magnified)

Materials: compound microscope, pre-made slide, lens tissue

1. Make sure your microscope lenses are clean. If they are dirty, clean them with the lens tissue. Do not clean them with anything except a lens tissue. Other materials might scratch the lenses.
2. Turn the objectives until the low power objective clicks into place.
3. Turn the coarse adjustment wheel until the low power objective just clears the stage.
4. While looking into the eyepiece, adjust the mirror or turn on the light to get the most amount of light possible.
5. Center the slide over the hole in the stage then move the clips to hold the slide in place.
6. While looking into the eyepiece, turn the coarse adjustment wheel toward you. This will raise the objective. Keep turning it until the slide comes into focus.
7. Use the fine adjustment wheel to sharpen the image.
8. Draw a picture of what you see.

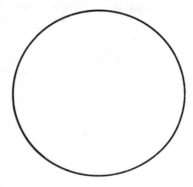

9. Switch to the high power objective. Use the fine adjustment wheel to sharpen the focus. (Never use the coarse adjustment wheel on high power.)
10. Draw a picture of what you see.

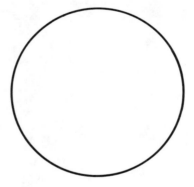

You will often use a microscope for experiments in this book. It can open your eyes to a whole new world. Review steps 1–7 above to learn how to take good care of the microscope.

Try this #3

Materials: microscope, glass slide, paper towel
1. Thoroughly wash and dry a glass slide.
2. Press your finger on the slide. This will leave your fingerprint.
3. Examine the slide under low power.
4. Draw your fingerprint as you see it under low power.

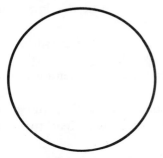

5. Switch to the high power objective.
6. Draw your fingerprint as you see it under high power.

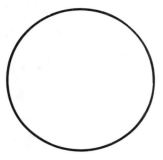

7. Compare your drawings with a friend's. Do they look the same? They shouldn't; no two people have the same fingerprints.

Materials: microscope, glass slide, scissors, paper towel, cover slip, lens tissue, eye dropper, water

1. Wash and dry the cover slip and the glass slide.
2. Cut out a small piece of lens tissue about the width of your pencil.
3. Place the piece of lens tissue on your slide.
4. Use the eye dropper and put one drop of water on the lens tissue.
5. Place the cover slip on the edge of the slide.

6. Gently lower the cover slip onto the drop of water.
7. Remove any bubbles by gently pressing on the cover slip with the end of your pencil.
8. Examine your slide under low power.
9. Draw a picture of what you see.

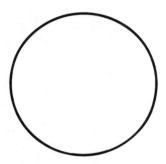

10. Examine your slide under high power.
11. Draw a picture of what you see.

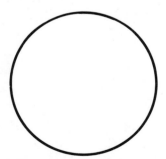

The type of slide you have just made is called a **wet mount**. A wet mount is a microscope slide that is made with water.

Chapter checkup

Fill in the blanks.

1. The magnifying power of a microscope can be found by multiplying the power of the _____ times the power of the _____.

2. A lens should be cleaned only with _____.

3. When using a microscope, the _____ power objective should always be used first.

4. The _____ adjustment wheel is used to sharpen the focus.

5. A slide that is made with water is called a _____.

cell
nucleus
cytoplasm
cell membrane
vacuole
chloroplasts
cell wall
cellulose

4. Cells

From far away, a jigsaw puzzle looks like a single picture. When you get close, though, you can see all the parts that make up the puzzle.

The onion skin is like a jigsaw puzzle. At first glance, it looks as if it is a solid layer of material. Only with the help of the microscope can we see all the tiny building blocks of which it is made.

These tiny building blocks are **cells**. A cell is the smallest living part of a plant or animal. Some living things are made up of many cells. Your body has over one trillion (1,000,000,000,000) cells. There are also living things that are made up of only one cell.

Try this #5

Materials: microscope, glass slide, cover slip, paper towel, iodine stain, onion, eye dropper
1. Put one drop of water on the glass slide.
2. Carefully peel off one thin layer from the inside of the onion.
3. Put the onion skin into the drop of water. Keep the onion skin as flat as possible.
4. Put the cover slip over the water drop.
5. Put one drop of iodine stain on the edge of the cover slip.
6. Dip the tip of the paper towel into the water at the other edge of the cover slip. This should draw the iodine onto the onion skin.
7. Observe the slide under low power.
8. Draw what you see.
9. Observe the slide under high power.
10. Draw what you see.

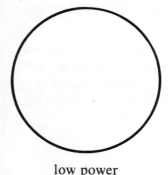
low power

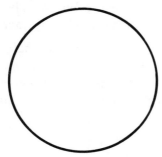
high power

The parts of a cell

Some cells are so small that millions of them could fit on the tip of your pencil. Other cells are much larger. Some cells are long and thin and others are round.

Cells have four important parts: a nucleus, cytoplasm, vacuoles, and a cell membrane. There are other smaller parts, too.

The **nucleus** controls all activities that go on in the cell. (There are some bacteria and algae that do not have nuclei.)

The **cytoplasm** is a jelly-like substance that surrounds the nucleus. The cytoplasm helps keep the cell alive.

Vacuoles are tiny sacks in the cell. They store liquids or food. Vacuoles are not always very easy to see in animal cells.

The **cell membrane** is a thin layer of material that holds the cell together. Everything that passes in or out of the cell must pass through the cell membrane.

Plant cells

Plant cells have a nucleus, cytoplasm, vacuoles, and a cell membrane. In many plant cells, the vacuoles are easy to see and seem to be the main part of the cell. Plant cells also have two other parts that animal cells do not have. Plant cells have chloroplasts and a cell wall.

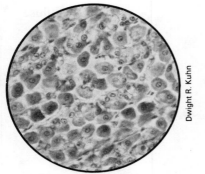

Cat stomach cells

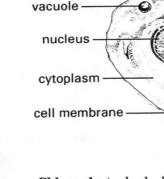

Animal Cell

vacuole

nucleus

cytoplasm

cell membrane

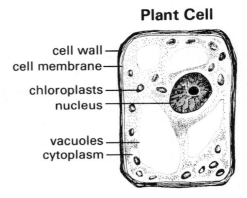

Plant Cell

cell wall
cell membrane
chloroplasts
nucleus

vacuoles
cytoplasm

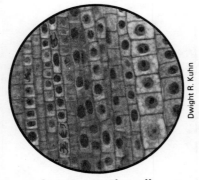

Onion root tip cell

Chloroplasts look like tiny green eggs in the cytoplasm. There is a substance in the chloroplasts that takes energy from sunlight and changes it into food for the cell. (We will talk more about that substance in chapter 14.)

A **cell wall** completely covers the cell membrane. It is made of a hard sticky material called **cellulose**. The hard cell walls help the plant stand up. The cellulose holds the cells together.

So, animal cells have four main parts: the nucleus, the cytoplasm, the vacuoles, and the cell membrane. Plant cells have six main parts: the nucleus, the cytoplasm, the vacuoles, the cell membrane, the chloroplasts, and the cell wall. Vacuoles usually show up most clearly in plant cells.

Try this #6

Materials: microscope, glass slide, cover slip, paper towels, iodine stain, toothpick

1. Put one drop of water on the glass side.
2. Take the toothpick and scrape the inside of your cheek.
3. Put the material you scraped from your cheek into the drop of water on the glass slide.
4. Put the cover slip over the drop of water.
5. Put one drop of iodine stain on the edge of the cover slip.
6. Dip the tip of the paper towel into the water at the other edge of the cover slip.
7. Observe the slide under low power.
8. Draw what you see.

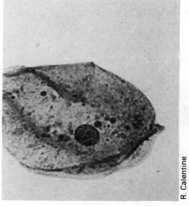

Human cheek cell

R. Calentine

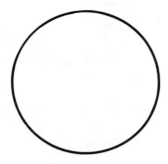

9. Observe the slide under high power.
10. Draw what you see.

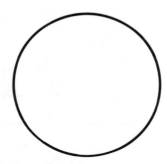

11. You have just looked at the cells that line the inside of your cheek. Do they have cell walls like the onion cells have? _____
12. Onion cells have a nucleus. Do cheek lining cells have a nucleus? _____

The onion cells you looked at had cell walls. An onion is a plant. Your cheek lining cells do not have cell walls. Cheek cells are animal cells and animal cells do not have cell walls.

Chapter checkup

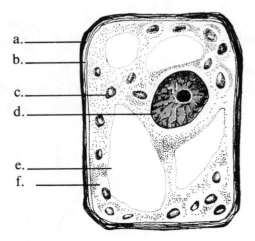

1. Label the diagram as a plant cell or an animal cell.
2. Label the parts of this cell. Use these words: nucleus, cytoplasm, vacuoles, cell membrane, cell wall, chloroplasts.

   a. _____

   b. _____

   c. _____

   d. _____

   e. _____

   f. _____

5. Cells and tissues

Look at the skin on your hand. Your skin is made up of millions of cells. All of these cells are alike. They all have the same job. A group of cells that look alike and do the same thing are called **tissue**.

Tissues are named by the kind of cells they have. Humans and many other animals have five different types of cells and tissues: bone, nerve, muscle, blood, and skin.

Bone tissue makes up your skeleton. The bone tissue gives your body its shape and helps support it.

Nerve tissue carries messages throughout your body. Cells called **neurons** make up nerve tissue. As you read this page, neurons are carrying messages up to your brain. Nerve tissue makes your muscles move. Nerve tissue controls your breathing and heartbeat.

Muscle tissue works with nerve tissue to help your body move. The nerve tissue sends messages from your brain to your muscle tissue. Your heart is mostly muscle tissue.

Blood cells in a liquid make up your blood tissue. Blood tissue carries food, water, and oxygen to your other tissues. Blood tissue helps your body fight germs.

Skin is made up of covering tissue. Covering tissue protects your body from things outside your body. Covering tissue also lines many things inside your body, such as your stomach.

Human skin
magnified 100 times

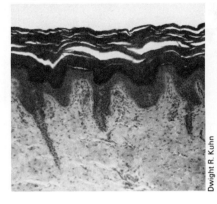

Dwight R. Kuhn

Ground bone cell
magnified 250 times

Dwight R. Kuhn

Human blood cells
magnified 1,000 times

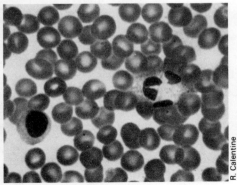

R. Calentine

Muscle tissue
magnified 400 times

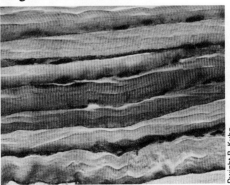

Dwight R. Kuhn

Nerve cell
magnified 250 times

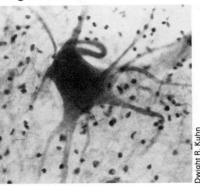

Dwight R. Kuhn

Try this #7

Materials: chicken wing, knife, paper towel
1. Put the chicken wing on a paper towel.
2. Examine the outside of the chicken wing.
3. What kind of tissue covers the outside of the chicken wing?

   _____

4. Remove the outside tissue.
5. What kind of tissue lies just beneath the outside tissue? _____
6. Try to locate a blood vessel. The blood vessel carries blood tissue.
7. Look for a long, white, stringy substance between the muscles. This is nerve tissue.
8. The hard tissue forming the skeleton is _____ tissue.

The tissue that is on the outside of the chicken wing is covering tissue. It is the skin of the chicken. Just below the skin is muscle tissue. The muscle is the meat of the chicken. The skeleton of the wing is bone tissue.

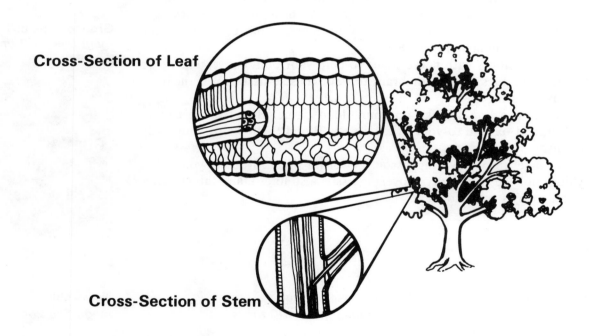

Cross-Section of Leaf

Cross-Section of Stem

Plant tissues

Plants are made up of tissues too. In the leaves there are tissues that make food. The chloroplasts in the cells combine sunlight, water, and carbon dioxide to make food.

Plants have special tissue to carry water. Water enters the roots of plants and travels up to the leaves through these tissues.

Plants also have protective tissue. These tissues shield the plant from harm. Tree bark is an example of protective tissue.

Cell reproduction

Cells grow and reproduce. That is, they make new cells. Single-cell plants and animals are organisms that are made up of only one cell. When they reproduce, they make an entire new organism. Many-celled plants and animals make new cells in order to grow. Tissues are made up of cells that have reproduced over and over again. Many-celled organisms need to make new cells to replace cells that have died.

Cells reproduce by a process called **mitosis**. Mitosis is cell division. One cell splits into two cells. The two cells look alike. We will look more at mitosis in chapter 27.

Try this #8

Materials: microscope, prepared slide of onion
1. Observe the onion slide under low power. Compare what you see with the illustrations above. Do you see any cells dividing? Draw what you see.

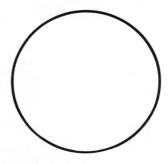

2. Observe the slide under high power. Draw what you see.

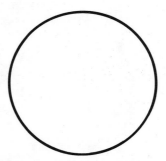

You probably saw cells in various stages of mitosis. Each cell divides and grows many, many times. That is what made the onion grow.

Chapter checkup

Write **T** for true or **F** for false in the blanks below.

_____ 1. Muscle tissue carries messages from your brain to other parts of your body.

_____ 2. Bark is the protective tissue on a tree.

_____ 3. A tissue is a group of different types of cells all working together.

_____ 4. Cells called neurons make up nerve tissue.

_____ 5. In the leaves of trees there are food-making tissues.

_____ 6. There are over one hundred different types of tissues in the human body.

_____ 7. Skin is an example of covering tissue.

_____ 8. Cells, except sex cells, divide in a process called mitosis.

Unit review

- Biology is the study of living things.
- Living things are made up of cells. A cell is the basic unit of life.
- Scientists say living things use energy, grow, reproduce, and respond to their environment (surroundings).
- An organism is a living thing.
- An organism can be made of one cell or many cells.
- A microscope is a device that makes things look larger than they are.
- Most cells have a nucleus, cytoplasm, cell membrane, and vacuoles.
- Plant cells have two extra parts that animal cells do not have: a cell wall and chloroplasts.
- A group of cells that look alike and do the same thing are called tissue.
- Cells, except sex cells, reproduce by mitosis, or cell division.

Unit checkup

True or False

_____ 1. A compound microscope has the same number of lenses as a simple microscope.

_____ 2. A slide made with water is called a wet mount.

_____ 3. All cells look alike.

_____ 4. The nucleus controls the activities of the cell.

_____ 5. Animal cells have chloroplasts.

_____ 6. A tissue is a group of cells that work together.

_____ 7. Plants have tissues, animals do not.

_____ 8. Cells cannot grow or reproduce.

_____ 9. An organism is a living thing.

_____ 10. All organisms are made up of millions of cells.

Matching

_____ 1. cell   a. a hand lens

_____ 2. mitosis   b. instrument with many lenses that makes objects look larger

_____ 3. microscope   c. slide made with water

_____ 4. cytoplasm   d. smallest living part of a plant or animal

_____ 5. chloroplast   e. plant cell part that helps plant make food

_____ 6. nucleus   f. cell part surrounding cell nucleus

_____ 7. vacuole   g. cell part that stores water and food

_____ 8. wet mount   h. cell division

_____ 9. tissue   i. group of like cells that do the same thing

_____ 10. magnifying glass   j. cell part that controls activities of cell.

Fill in the answers

A. Name the parts of an animal cell.

     1._____

     2._____

     3._____

     4._____.

B. Name the parts of a plant cell.

     1._____

     2._____

     3._____

     4._____

     5._____

     6._____

Unit 2.
Taxonomy: where does it fit?

More than one and a half million organisms are known to exist in the world. How can so many organisms be studied? Scientists find it easier to study organisms when they are put into groups. Organisms with many similarities are put into the same group. In this unit you will learn how we group organisms and the important features of each group.

Jan Doyle

Bev Hehkop   Unicorn Stock Photos

R. Calentine

Barbara Durham

Stephen Thompson

classify
taxonomy
kingdom
protist
euglena
genus
species

6. How organisms are classified

To **classify** means to group things together according to features they have in common. Suppose you went to a hardware store to buy a shovel. But there was no order to the store. Everything was piled in a heap. It would be very difficult to find a shovel or anything else.

Lucky for us, stores use a system of classification. Stores usually classify their stock in sections. The items in each section have something in common. In a hardware store there is the garden supply area, the plumbing area, and various other areas. The shovels in an area may even be classified into different types of shovels. Classification gives order to the store.

Taxonomy

Scientists classify all organisms. **Taxonomy** is the scientific classification of organisms. The first taxonomist was a Greek man named Aristotle. He classified all living organisms into plants and animals. He divided the plants into herbs, shrubs, and trees. He divided the animals into land animals, water animals, and air animals.

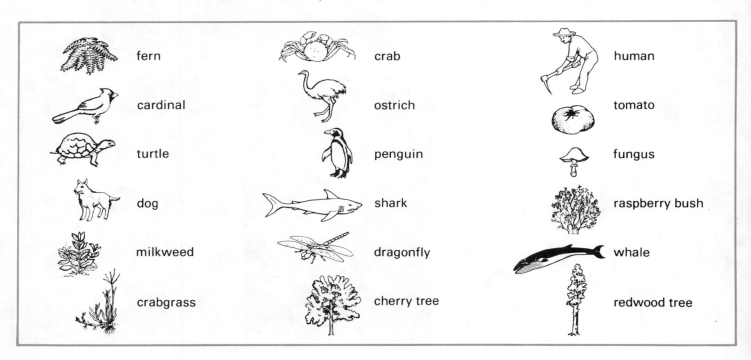

fern	crab	human
cardinal	ostrich	tomato
turtle	penguin	fungus
dog	shark	raspberry bush
milkweed	dragonfly	whale
crabgrass	cherry tree	redwood tree

Things to do

Suppose you are the world's first taxonomist. Your job is to classify the organisms listed on page 30. Put them into either Group A or Group B according to something they have in common. Give each group a title. Then divide the organisms in Group A into two smaller groups. Give these groups a title. Divide the organisms in Group B into two smaller groups. Give these groups a title.

Group A _____   Group B _____
           (Title)   (Title)

_____   _____

_____   _____

_____   _____

_____   _____

_____   _____

_____   _____

_____   _____

_____   _____

Divide Group A into two smaller groups. Give these groups a title. Do the same for Group B. You will not need to fill in each blank.

Group A   Group B

_____   _____   _____   _____
(title)   (title)   (title)   (title)

_____   _____   _____   _____

_____   _____   _____   _____

_____   _____   _____   _____

_____   _____   _____   _____

_____   _____   _____   _____

Now share your classifications with other members of your class. Did you have the same classifications? Explain why you chose to classify the organisms as you did. Ask your classmates to explain their classifications. Think about the job scientists have classifying all the plants and animals in the world.

Kingdoms

When you classified the organisms, your list probably did not look like your neighbors? Did everyone in the class agree? Probably not. Classifying organisms can be very difficult. Even today not all scientists agree on exactly how each organism should be classified. Most scientists, however, do use the same general system. That system divides all organisms into three major groups. These groups are called **kingdoms**. There is the animal kingdom, the plant kingdom, and the protist kingdom.

Birds, insects, reptiles, and mammals are all in the animal kingdom. Trees, flowers, grass, and shrubs are all in the plant kingdom. All plants are able to make their own food.

The third kingdom is the **protist** kingdom. Protists are one-celled organisms. They are neither definitely animal nor definitely plant. Most protists have traits of both animals and plants. Some protists have traits completely different from animals and plants. The **euglena** is a protist. It looks and moves about like an animal. It makes its own food like a plant. Yeast is a protist too.

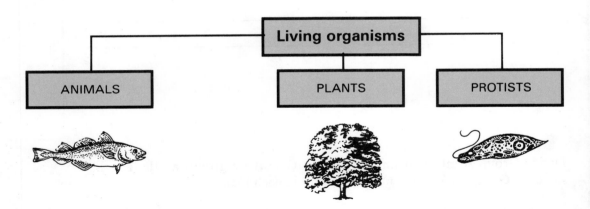

In *Things to do*, you divided the organisms into two large groups, according to certain characteristics. Then you took each group and tried dividing it again. Scientists do the same thing. But they use seven classifications. These are the seven:

Kingdom
Phylum
Class
Order
Family
Genus
Species

Each division has fewer members than the one above it. The members in each smaller division share more characteristics.

Look at the chart below. It shows the classification of humans. Notice how each group becomes more specific. The last classification, species, has only one kind of living thing in it.

Classification of humans   **Traits**

Kingdom
  Animal . . . . living things that are not plants

Phyllum
  Chordata . . . . at some time in its life has a cord that runs along its body; some animals in this group have a backbone

Class
  Mammal . . . . has a backbone, hair, produces milk for its young

Order
  Primate . . . . has a backbone, hair, produces milk for its young, stands upright, has fingers and thumbs

Family
  Hominidae . . . . has backbone, hair, produces milk for its young, has fingers and thumbs, stands upright, walks upright

Genus
  Homo . . . . has backbone, hair, produces milk for its young, has fingers and thumbs, stands upright, walks upright, has advanced brain

Species
  Sapiens . . . . has backbone, hair, produces milk for its young, has fingers and thumbs, stands upright, walks upright, has a more advanced brain

Homo sapiens (human)

Lepomis macrochirus (bluegill sunfish)

Orchelimum vulgare (grasshopper)

Felis domesticus (house cat)

Scientific names

Many animals have several common names. The crapper and calico bass are names for the same fish. To avoid confusion, scientists have developed a special system. With this system no two animals have the same scientific name. Each organism is given a first and last name. The first name is the **genus** name. The last name is the **species** name. The scientific name for modern humans is *Homo sapiens*. The scientific name for a grasshopper is *Orchelimum vulgare*. The scientific name for a house cat is *Felis domesticus*. The scientific name for the blue gill sunfish is *Lepomis macrochirus*.

33

Chapter checkup

Fill in the blanks below:

1. _____ is the scientific classification of living organisms.

2. The _____ is an organism in the protist kingdom.

3. The _____ and _____ names are used when a scientific name is given to organisms.

4. To group things by something they have in common is to _____ them.

5. The three major groups into which all organisms are divided are called _____ .

6. The kingdom that includes one-celled organisms that are neither plant nor animal is the _____ kingdom.

reptiles	spiny-skinned
fish	hollow-bodied
amphibians	worms
arthropods	sponges
mollusks	vertebrae
mammals	vertebrates
birds	invertebrate

7. Classification of animals

Gail Denham

Dwight R. Kuhn

You are probably most familiar with the animal kingdom because you are a part of it.

Vertebrates and invertebrates

Scientists divide animals into two large groups:
1. Animals with a backbone
2. Animals without a backbone

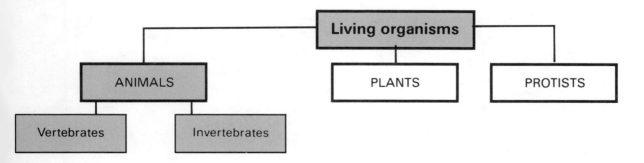

Try this #9

Materials: chicken neck, pan, hot plate, vinegar, tweezers, spoon
1. Fill the pan with water.
2. Add one spoonful of vinegar to the water.
3. Put the chicken neck in the solution and boil it for about 15 minutes.
4. Carefully take the chicken neck out of the water and put it on a paper towel.
5. Use the tweezers to remove all the meat from the bones.
6. Carefully examine the bones.

The bones you examined are called **vertebrae**. The vertebrae are part of the backbone. Any animal with a backbone is called a **vertebrate**. A chicken is a vertebrate. Dogs, birds, snakes, toads, and fish are vertebrates. You have a backbone. You are a vertebrate.

Animals without backbones are called **invertebrates**. Worms, snails, and crabs are all invertebrates.

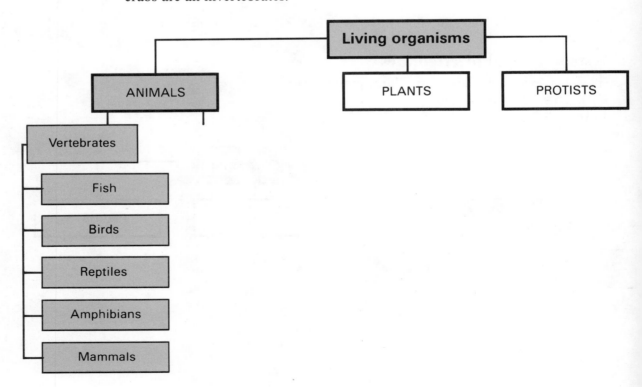

Snake (reptile)

Bullfrog (amphibian)

Saw-whet owl (bird)

Vertebrates

Vertebrates are divided into five classes. The animals in each class have something in common with one another.

Mammals are the only animals with hair. Bears and mice have thick hair called fur. Elephants and armadillos have very little hair. Most mammals do not lay eggs. The young develop inside the mother. Mammals are the only animals that produce milk for their young. Mammals are warm-blooded. Warm-blooded means the animals' blood stays a certain temperature all the time.

Birds are the only animals with feathers. Most birds can fly. To fly, their bodies must be light. Birds have hollow bones to make their bodies light. Birds lay eggs. Like mammals, birds are warm-blooded.

Reptiles have scales on their bodies. Snakes, lizards, turtles, and alligators are all reptiles. Most reptiles lay eggs. Reptiles are cold-blooded. A cold-blooded animal cannot control its body temperature. Most cold-blooded animals sun themselves to get warm. They go to water or shade to cool down.

Fish live in water. All fish have gills. Gills take the oxygen out of water. A fish uses its gills to breathe as we use our lungs. Most fish have scales on their bodies. Most fish lay eggs. Like reptiles, fish are cold-blooded.

Amphibians are born in the water. Like fish, they breathe with gills. As they grow older they lose their gills. They develop lungs and live on land. Frogs, toads, and salamanders are all amphibians. Like reptiles and fish, amphibians lay eggs and are cold-blooded.

Bette fish (fish)

Prairie dog (mammal)

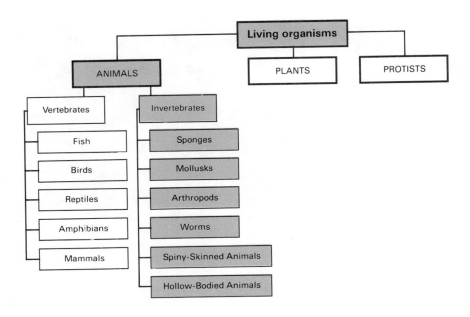

Invertebrates

Invertebrates do not have backbones. In fact, invertebrates have no bones at all. Some invertebrates, like the grasshopper and the crab, have hard coverings that give their body shape. Invertebrates like the worm and slug have soft bodies with no covering. Invertebrates are divided into six classes:

Arthropods are invertebrates with jointed legs. The word *arthro* means "joint." *Pod* means "leg." Arthropods are the largest class of animals. In fact, there are more arthropods than there are all other animals combined. Insects and spiders are arthropods. Crabs, shrimp, and lobsters are arthropods.

Mollusks have soft bodies. Most mollusks have shells to protect their bodies. Clams, oysters, and snails are mollusks with shells. Squids and octopuses are mollusks without shells.

Spiny-skinned animals have spines that cover almost their entire bodies. They also have tube feet that end in suckers. Sea urchins and starfish are spiny-skinned animals.

Hollow-bodied animals have soft, baglike bodies. They all live in the water. Jellyfish and sea anemones are hollow-bodied animals.

Worms are long soft-bodied creatures with no legs. Earthworms, tapeworms, and roundworms are all in the worm class.

Sponges have a tough flexible body with many holes. All sponges live in water. The holes take in water. They filter small bits of food from the water.

Sponge

Snail with eggs (a mollusk)

Garden spider (an arthropod)

Earthworm

Chapter checkup

Write the letter in the space provided.

A. Matching

_____ 1. Animals with hollow bones

_____ 2. Animals born with gills, but develop lungs as adults

_____ 3. Turtles and alligators are in this class

_____ 4. Animals with backbones

_____ 5. Live their entire lives in water

_____ 6. Produce milk for their young

a. amphibians
b. vertebrates
c. fish
d. birds
e. mammals
f. reptiles

B. Matching

_____ 1. Animals with soft, baglike bodies

_____ 2. Animals with jointed legs

_____ 3. Oysters and snails are in this class

_____ 4. Animals with no backbones

_____ 5. Animals with many-holed bodies

_____ 6. Soft-bodied creatures with no legs

_____ 7. Animals with tube feet that end in suckers

a. sponges
b. hollow-bodied
c. mollusks
d. invertebrates
e. spiny-skinned
f. worms
g. arthropods

botanist
algae
mosses
liverworts
ferns
seed plants
spores

8. Classification of plants

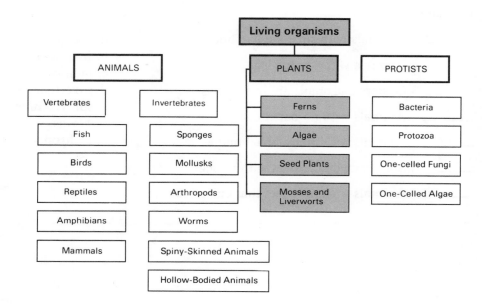

Seaweed is an alga. Algae are the simplest kind of plants.

R. Calentine

Botanists are scientists who study plants. Botanists disagree about how to classify plants. But many divide the plant kingdom into four phyla (the plural of phylum): **algae** (some kinds), **mosses** and **liverworts**, **ferns**, and **seed plants**.

Algae

Scientists call some kinds of algae plants. (Other kinds are protists. We'll look at them in the next chapter.) Red, brown, and green algae are in the plant kingdom. Most algae are water plants. Some algae are one-celled and can only be seen with a microscope. Others are easily seen with the naked eye. Seaweed is an alga. The green "scum" on the surface of ponds is algae too.

Algae are the simplest kind of plant. Algae have chloroplasts so they can make their own food. But they have no true roots, stems, or leaves. They do not have different types of tissue. Even an alga of many cells has only one kind of cell.

Mosses are usually found in damp, shady places.

Liverworts get their name from their leaflike parts that look like livers.

Mosses and liverworts

Mosses look like small clumps of green carpet. What is commonly called a moss is actually many tiny moss plants growing together. There are about 24,000 species of mosses and liverworts.

Liverworts are small green plants with flat leaves that lie flat to the ground. They are usually found near the banks of streams or ponds.

Mosses and liverworts have no true roots, stems, or leaves. All their cells are almost alike. Water moves from cell to cell through the cell walls. It is a very slow process. To get their much needed water, mosses and liverworts grow very close to the ground. They must live in moist, shady places.

Mosses and liverworts have sexual and asexual stages of reproduction. Sexual reproduction involves the joining of a male and a female cell. Asexual reproduction involves the dividing of one cell only. In mosses and liverworts the asexual stage produces **spores.** Spores are tiny cells that fall from a plant. They grow into new plants. The new plants produce male and female sex cells that unite to form yet another new plant. These stages of sexual and asexual reproduction will repeat themselves again and again.

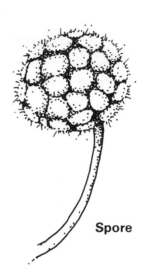

Spore

Ferns

Ferns, like mosses and liverworts, tend to grow in moist, shady places. Most ferns are about two feet tall. Some ferns are very tiny and look like mosses. Other ferns are as large as trees. There are about 12,000 species of ferns.

Ferns are more complex than mosses or liverworts. They have many different types of cells. They also have veins that carry food and water up the stems to the leaves.

Ferns have sexual and asexual reproductive stages. During one stage, the ferns, like mosses and liverworts, produce spores. The spores look like small brown dots on the leaves. These fall from the plant. They produce male and female sex cells. The male and female cells join. A new fern grows. In time that fern will produce spores and the cycle will repeat.

Ferns have roots, stems, and leaves.

Seed plants

Trees and flowers are seed plants. Fruits and vegetables come from seed plants. Seed plants are the most dominant plants on earth today. There are hundreds of thousands of species. Seed plants are the most complex plants. Like the ferns, seed plants have many different types of cells. Like the ferns they have veins that carry food and water. But, unlike the ferns, seed plants do not reproduce by spores. They reproduce by seeds. Male and female sex cells join and produce a seed. Unlike a spore, a seed has a shell that protects the plant developing inside it.

Reproduction by seed is more efficient than reproduction by spore. Only a few of the many spores that fall develop into plants. Few get the right mix of light and water. A seed, however, has food inside it for the tiny plant and a protective shell.

Trees reproduce by seeds.

Dandelions reproduce by seeds. Wind helps scatter the seeds.

Things to do:

Seeds are scattered in many different ways. Some are blown about by the wind. Some attach themselves to animals. Some are carried by water. Some are shot out by the mother plant which acts like a cannon. Sometimes humans spread seeds for their own benefit. Look at the pictures of the seeds below. Tell how you think each is scattered.

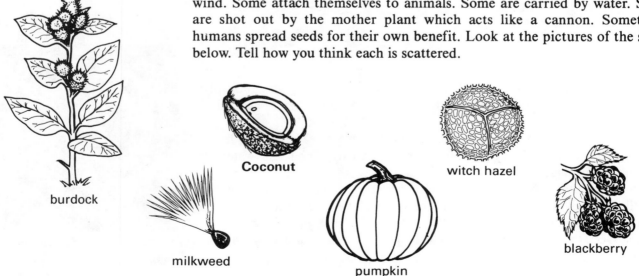

burdock

milkweed

Coconut

pumpkin

witch hazel

blackberry

Chapter checkup

Write **T** for true or **F** for false.

_____ 1. Plants are divided into algae (some kinds), mosses and liverworts, ferns, and seed plants.

_____ 2. Mosses reproduce by seeds.

_____ 3. There are many different types of cells in a fern.

_____ 4. Liverworts carry water to their leaves in veins.

_____ 5. Ferns reproduce by spores.

_____ 6. Most of the plants we see around us are ferns.

_____ 7. Botanists are people who study animals.

_____ 8. Mosses grow up to 30 feet tall.

_____ 9. Seeds are the most efficient way a plant can reproduce.

_____ 10. Ferns have both a sexual and asexual reproductive stage.

protist
bacteria
fungi
algae
protozoa
ameba
euglena
paramecium

9. Classification of protists

Things to do

How would you divide these figures into two groups according to their shapes?

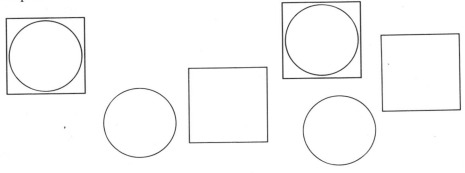

Now divide those same figures into *three* groups according to their shapes.

In the first exercise the squares should be put into one group. The circles should be put into the other. But what about the other three figures? They have traits of both squares and circles. What did you do with these "in-between" figures?

The second time you classified the figures was probably easier. There was a place to put these "in-between" figures.

When classifying organisms, scientists have a similar problem. Some organisms are obviously plants. Some organisms are obviously animals. But there are also "in-between" organisms. These organisms have both plant and animal traits.

To solve this classification problem scientists developed a third kingdom. They call it the **protist** kingdom. Protists are organisms that have characteristics of both plants and animals.

There are four major groups in the protist kingdom: **bacteria**, some **fungi**, some **algae**, and **protozoa**.

The organisms in the protist kingdom are single-celled organisms. A single-celled organism is an organism made of only one cell. Inside that one cell all the needed life processes go on—eating, digestion, waste removal, and reproduction.

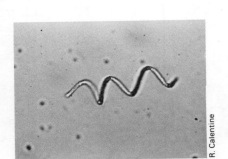

Spirillum bacteria (spiral shaped)

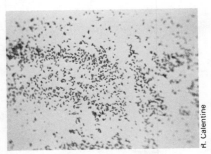

Coccus bacteria (round)

44

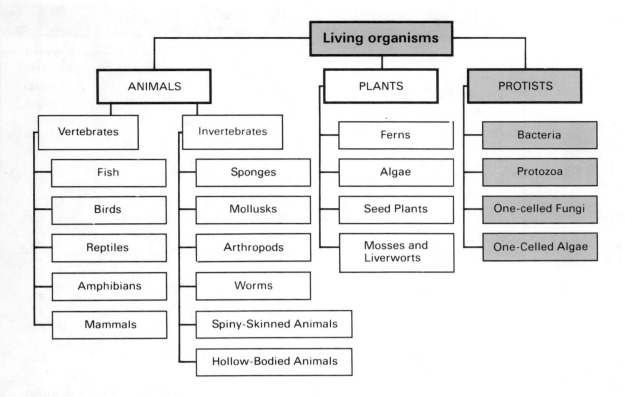

```
                        ┌─────────────────┐
                        │ Living organisms │
                        └─────────────────┘
        ┌───────────────────────┼───────────────────────┐
 ┌─────────────┐        ┌─────────────┐          ┌─────────────┐
 │   ANIMALS   │        │    PLANTS   │          │   PROTISTS  │
 └─────────────┘        └─────────────┘          └─────────────┘
```

ANIMALS		PLANTS	PROTISTS
Vertebrates	Invertebrates	Ferns	Bacteria
Fish	Sponges	Algae	Protozoa
Birds	Mollusks	Seed Plants	One-celled Fungi
Reptiles	Arthropods	Mosses and Liverworts	One-Celled Algae
Amphibians	Worms		
Mammals	Spiny-Skinned Animals		
	Hollow-Bodied Animals		

Bacteria

Bacteria are very tiny single-cell organisms. They are smaller than most other cells. They are so small they cannot be seen with a classroom microscope. They can be seen only with an electron microscope. An electron microscope magnifies objects up to 300,000 times.

Bacteria come in three basic shapes. Some bacteria are shaped like rods. Some are shaped like corkscrews. Others are round.

Many bacteria are helpful to us. Plants could not grow without bacteria in the soil. Some bacteria work inside our bodies and make vitamins. There are bacteria in our stomachs that help digest food.

Some bacteria are harmful to us. They can cause diseases. They enter our bodies through the nose or mouth, or through an open cut. Then they reproduce. An illness caused by bacteria is called an infection.

Fungi

Fungi grow in many shapes and sizes. Yeast is one type of fungus. It is single-celled. But some fungi are many-celled. Molds and mushrooms are fungi with many cells. So they are a problem for taxonomists to classify. For that reason, many taxonomists place all fungi in their own kingdom. In this book, we are classifying only one-celled fungi as protists.

Most fungi look like plants. There is one important difference. Fungi cannot make their own food. Most fungi live on the remains of dead plants or animals. There are even types of fungi that feed on living plants and animals.

Fungi help us by decomposing plant and animal remains (making them decay or rot). This keeps our soil fertile. They help in making food products such as bread and cheese. Penicillin and other medicines are made from fungi. But fungi can also be harmful. They make food spoil. They also may cause many plant diseases.

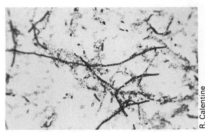

Bacillus bacteria (rod shaped)

Puffball fungi

Materials: baby-food jar with lid, bread
1. Rub the piece of bread on the floor.
2. Put the piece of bread in the baby-food jar.
3. Add three drops of water to the bread.
4. Cover the jar and store in a dark place for three to four days.
5. Examine the bread everyday for three or four days.
6. Describe what you see growing on the bread.

Mold was growing on the bread. Mold is a fungus. It is using the bread for food.

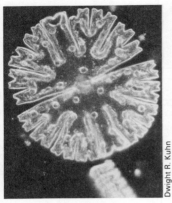

Alga cell

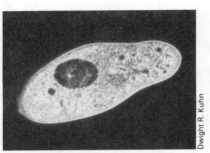

Paramecium *magnified 100 times*

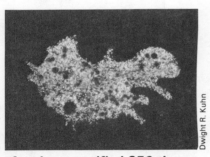

Ameba *magnified 250 times*

Euglena *magnified 40 times*

Algae

Blue-green algae are one-celled organisms. They live in water. They make their own food, just as plants do.

Algae are very important. They are the main food source for water life. Algae (both plant and protist types) produce most of the oxygen found in the water. Without algae most water life could not live.

Protozoa

There are more than 15,000 different species of protozoa. Protozoa are sometimes called one-celled animals. Hundreds of small, wriggling protozoa live in a single drop of pond water. Some protozoa are shapeless blobs. Others have definite shapes. Some have small hairlike structures on their sides. Some appear to have tails.

Three of the most common protozoa are the **ameba**, the **euglena**, and the **paramecium**.

The ameba looks like a tiny drop of jelly. It is protected by a thin outer membrane. Inside is cytoplasm and a nucleus. It moves about very slowly. One side extends out and the rest of the ameba follows it. Amebas help clean up ponds and streams by eating small bits of dead plants or animals. They also serve as food for small fish.

The euglena is a small thin organism with a threadlike tail. By moving its tail back and forth it can move about in water. The euglena has chloroplasts like plants do. So it can make its own food. It can also absorb food through its outer membrane.

The paramecium is one of the most complex single-celled organisms. It looks like a slipper with tiny hairs along its sides. It uses the hairs like tiny oars to move itself. Compared to an ameba, a paramecium moves very quickly.

Try this #11

Materials: microscope, eye dropper, pond water, glass slide, cover slip
1. Using the eye dropper, put one drop of pond water on a glass slide.
2. Place a cover slip over the pond water.
3. Examine the slide under the microscope.
4. Draw pictures of the different organisms you see.
5. Look at the pictures of the ameba, the euglena, and the paramecium.

Were any of these in the water sample you looked at? _____

Were there algae in the drop of water? _____

Chapter checkup

Fill in the blanks below with the following words:

oxygen   infections
fungi   algae
paramecium   protists
electron microscope   penicillin
ameba

1. A jelly-like protozoan is the _____.

2. _____ are organisms that are neither definitely plant nor animal.

3. Algae produce most of the _____ in the water.

4. Yeasts and mushrooms are two types of _____.

5. Illnesses caused by bacteria are called _____.

6. The _____ is one of the most complex single-celled organisms.

7. _____ and other medicines are made from fungi.

8. Bacteria are so small they can only be seen with an _____

    _____.

9. Some _____ are considered plants and others are considered protists.

Unit review

- Organisms are classified according to similarities or differences.
- When classifying organisms, each subdivision is more specific than the one before it.
- Organisms can be divided into three kingdoms: animal, plant, and protist.
- Animals are divided into two phyla: vertebrates and invertebrates.
- The five classes of vertebrates are mammals, birds, reptiles, fish, and amphibians.
- The six classes of invertebrates are arthropods, mollusks, spiny-skinned animals, hollow-bodied animals, worms, and sponges.
- Plants are divided into four major groups—algae(multi-celled variety), mosses and liverworts, ferns, and seed plants.
- Mosses, liverworts and ferns produce both sexually and asexually.
- Seed plants are the highest plant form. They have veins, specialized cells, and reproduce by seed.
- Protists are one-celled organisms that are not clearly plant or animal.
- There are four major protist groups: bacteria, one-celled fungi, blue-green algae, and protozoa.

Fill in the chart   Use the unit review above to help you.

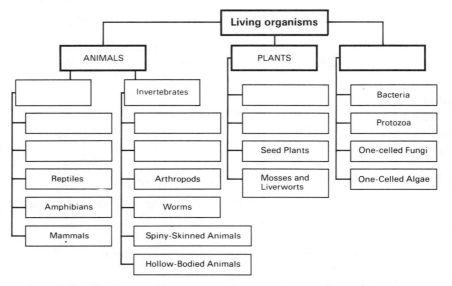

Unit checkup

Finish the sentence

1. The scientific classification of organisms is called _____ .
   (taxonomy, biology)

2. A(n) _____ is neither a plant nor an animal.
   (protist, invertebrate)

3. Ferns, mosses and liverworts reproduce by _____ .
   (spores, seeds)

4. An animal that has a backbone is a _____ .
   (vertebrate, invertebrate)

5. A _____ animal cannot control its body temperature.
   (cold-blooded, warm-blooded).

Unit 3.
Plant life

Plants are alive. They grow. Some live only a few hours. Others live for hundreds of years. Sunshine is vital to all plants. Plants use the sun's energy to make food. Plants also breathe. Breathing helps the plants make energy. Plants respond to things around them. They react to light, touch, heat and cold. They even respond to sounds.

We can learn a lot about life through the study of plants. They are a beautiful and complex part of life on Earth.

USDA

A. Franklin

James A. Anderson

Brian Parker Tom Stack & Associates

Thomas Fletcher

root
stem
leaves
flowers
seeds
fruit

10. Plants

Everywhere you look there are plants. Plants grow in the dry, hot desert and in the freezing arctic. They grow in valleys and on mountain tops. They grow in gardens and inside homes.

Plants are important to us

We could not live without plants. Plants supply humans and other animals with oxygen to breathe. Plants take in a gas called carbon dioxide. They give off oxygen. We breathe the oxygen and change it back to carbon dioxide. Plants and animals depend on each other for life.

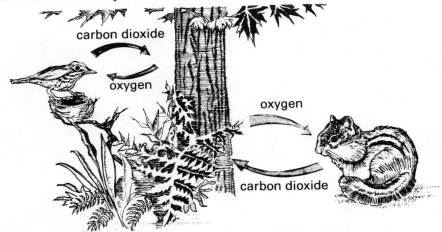

Plants take in carbon dioxide and give off oxygen. Animals take in oxygen and give off carbon dixoide.

All of our food comes from plants or from animals that eat plants. We eat the seeds of corn and rice. We eat the roots of carrots and beets. The leaves of cabbage and lettuce are food for us. Stems are the food in rhubarb and celery. We eat the flowers of broccoli and cauliflower. We eat the fruit of apple and orange trees. Grains and grasses are food for the cattle and other animals that many of us eat.

Many of the things we use every day come from plants. The wood we build with comes from trees. Trees also give us products such as paper, rubber, cork, and turpentine. Fuels such as coal and natural gas are the remains of decayed plants. Cotton plants give us material to make clothes. Many medicines come from plants.

Plants are used to make many products.

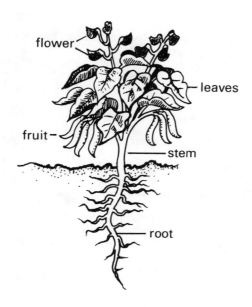

Parts of plants

Most plants have three major parts: the root, the stem, and the leaves. Some plants also have flowers and fruit. Each part has a special job. The **root** takes in water and minerals from the soil. The **stem** moves water and minerals up from the roots to the leaves. It moves the food made by the leaves down to the roots. The stem also helps support the plant. The **leaves** collect energy from the sun. They change this energy to food for the plant. The **flower** contains the reproductive parts of some plants. The **seeds** are made in the flower. The seeds are stored in the **fruit**. The fruit protects the seeds. It stores extra food made by the plant.

Try this #12

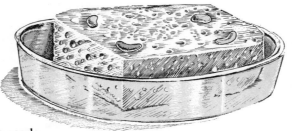

Materials: dish, sponge, four bean seeds
1. Put the sponge on the dish.
2. Fill the sponge with water.
3. Put the four bean seeds on the sponge near the edge.
4. Wait several days until the bean seeds sprout.
5. Even the young sprout will have the three major parts of a plant: root, stem, and leaves. Draw a picture of the sprout and label these three parts.
6. Save these sprouts, you will use them again in chapter 12.

Chapter checkup

Write **T** for true or **F** for false next to each statement below.

_____ 1. The leaves of a plant take in water and minerals.

_____ 2. Fuels such as coal and natural gas come from decayed plants.

_____ 3. Plants give off oxygen.

_____ 4. Water and minerals are carried through the stem up to the leaves.

_____ 5. All plants have flowers and fruit.

_____ 6. Seeds are stored in the stems of plants.

_____ 7. The stem helps support the plant.

_____ 8. The fruit stores extra food made by the plant.

fibrous root
taproot
epidermis
root hairs
cortex
central core
xylem cells
phloem cells

11. Roots

When you look at most plants you don't see the roots. The roots are below the ground. They act like sponges. They absorb water and minerals for the plant. Plants need water and minerals to live. Some roots even store food for the plant. They also help anchor the plant to the ground.

Grass has a fibrous root system

Try this #13

Materials: grass root system, dandelion root system
1. Carefully examine the grass root system.
2. Draw a picture of the grass root system.
3. Carefully examine the dandelion root system.
4. Draw a picture of the dandelion root system.
5. What do the root systems have in common? _____
_____
6. What is the major difference between the two root systems?
_____

Root systems

The grass you looked at had a **fibrous root** system. Fibrous roots have many different roots growing in all directions. These roots do not grow deep. They spread out near the ground's surface. They help hold the ground together. This helps prevent soil erosion. (Erosion is the gradual wearing away of the soil.)

The dandelion had one major root. This is called a **taproot**. Most taproots grow deeper than fibrous roots. Sometimes taproots store extra food made by the plant. We eat the taproots of vegetables such as carrots and beets.

Dandelions have one major root called a taproot.

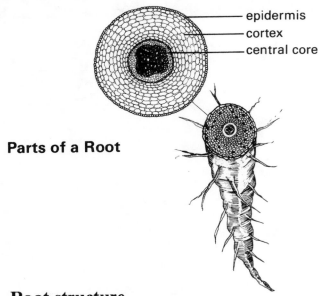

Parts of a Root

Root structure

Roots are divided into three main sections. The outermost part is the root's epidermis. Inside that is the **cortex**. The middle of the root is the **central core**.

An **epidermis** covers the outside of the root. It takes in water and minerals. It helps protect the root.

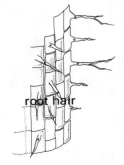

root hair

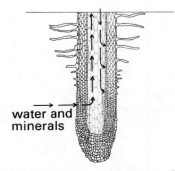

water and minerals

Root hairs take in water and minerals from the soil.

The fuzzy growth on the root is the **root hairs**. The root hairs are part of the epidermis. Each root hair is the extension of a single epidermis cell. The root hairs grow out into the soil. They help take in water and minerals.

Inside the epidermis is the **cortex**. The cortex stores excess food made by the plant. It also stores excess water taken in by the plant.

Inside the cortex is the **central core**. The central core runs up the middle of the root. In the central core, there are **xylem** and **phloem cells**. There are xylem and phloem cells throughout the plant, from the roots to the leaves. Both xylem and phloem cells have tiny holes in them. Materials flow from cell to cell through these holes.

The xylem cells take water and dissolved minerals up through the root and stem to the leaves. The leaves use the water and minerals to help make food for the plant. The plant uses this food for growth. Sometimes, the plant makes more food than it can use right away.

The phloem cells take this excess food down through the stem to the roots. That excess food is stored in the cortex until the plant needs it.

Try this #15

Materials: carrot slice
1. Carefully examine the carrot slice.
2. Draw a picture of the carrot slice.
3. Label the epidermis, cortex, and central core.

Chapter checkup

Fill in the blank in each question. Choose one of the two words written after the answer blank.

1. The _____ (cortex, central core) contains both xylem and phloem cells.

2. _____ (Xylem, Phloem) cells carry water and minerals up through the stem to the leaves.

3. _____ (Grasses, Dandelions) are plants with taproots.

4. The _____ (phloem, cortex) stores food for the plant.

5. _____ (Xylem, Phloem) cells carry food made by the leaves down to the root.

6. Roots that grow close to the surface are _____. (fibrous roots, taproots)

7. The _____ (epidermis, cortex) covers the outside of the root.

8. _____ (Grasses, Carrots) have fibrous roots.

9. _____ (Root hairs, Taproots) are epidermis cells that extend out into the soil.

herbaceous stems
dicot
monocot
chlorophyll
vascular bundles
pith
rind
woody stems
cambium
sapwood

12. Stems

The stem of a plant connects the roots to the leaves. It is like a pipeline. Xylem cells in the stem carry the water and minerals from the roots to the leaves. Phloem cells carry the food made in the leaves down to the roots. The stem also supports the leaves. Leaves need sunlight to produce food. The stem holds the leaves up to the sunlight.

Stems grow in two directions—width and length. This growth occurs because of the growth of new cells.

Try this #16

Materials: glass jar, food coloring, celery stalk, knife
1. Cut about one-half inch off the bottom of the celery stalk.
2. Put about one cup of water into the glass jar.
3. Add 10 drops of food coloring to the water.
4. Put the celery stalk in the water with the cut side down.
5. Let the celery stand for about one-half hour.
6. What do you observe happening to the celery stalk?

_____

What you saw were the xylem cells carrying the colored water up the stalk. If you wait long enough, you will see the colored water go up into the leaves.

Materials: geranium plant
1. Place the geranium plant near a window that receives lots of sunlight.
2. Notice the position of the leaves.
3. Let the plant stand for several days.
4. Again note the position of the leaves.
5. What happened to the leaves? _____

_____

If you look carefully you will see that most of the leaves are now facing the sun. Stems are able to slowly twist and turn. This stem turned to let the leaves receive more sunlight.

The herbaceous stem

Herbaceous stems are soft and green. Most small plants have herbaceous stems. The celery and geranium you used in the exercises above have herbaceous stems. Two orders of plants, **dicots** and **monocots**, have herbaceous stems.

The dicot's stem has a thin outside wall called an epidermis. The epidermis protects the stem. Inside is the cortex of the stem. The cortex cells contain **chlorophyll**. Chlorophyll is the substance that makes the plant green. It is also the substance that allows the plant to make its own food. So the cortex cells help the leaves produce food for the plant. These cells are also storage areas for the food.

Inside the cortex are the **vascular bundles**. The vascular bundles contain the xylem and phloem cells. In most dicot stems the vascular bundles form a ring around the **pith**. The pith is the center of the stem. The pith cells store food.

The monocot's stem has a tough outside covering called a **rind**. The rind protects the plant. It also helps support the plant. Inside the rind is the pith. Scattered about in the pith are the vascular bundles. The pith and the vascular bundles serve the same purpose in both the dicot and the monocot stems.

Dicot Stem

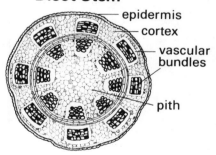

epidermis
cortex
vascular bundles
pith

Monocot Stem

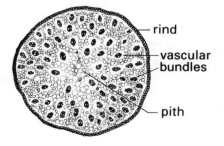

rind
vascular bundles
pith

The woody stem

All trees have **woody stems**. Woody stems are harder than herbaceous stems. They are not green because their cells do not contain chlorophyll. The outside of a woody stem is covered with bark.

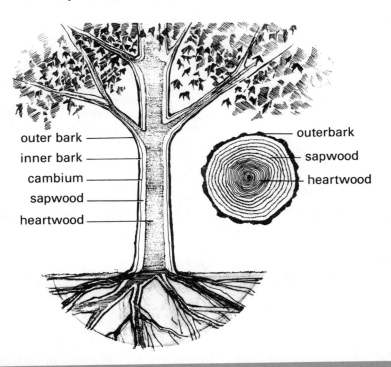

Try this #18

Material: tree branch
1. Rub your finger over the outside of the bark.

2. Describe the texture. _____

3. Peel some of the bark off the branch. Feel the inside of the bark.

4. Describe the texture. _____

5. One side of the bark has cells that are alive. The other side is covered with dead cells. Which side, the inside or the outside, do you think has the dead cells?

_____

The rough outer bark is covered with dead cells. The outer bark protects the tree from insects and diseases. The smooth inner bark is made of living cells.

The **cambium** makes the tree grow in width. Each year, the cambium produces a new layer of cells. The outermost cells form a new layer of bark. The food-carrying phloem cells are located here. The innermost cells form a new layer of wood. The water-carrying xylem cells are located here. This new layer of wood is called the **sapwood**. After many years, the cells in the sapwood die. These dead cells form the heartwood. The heartwood is the innermost part of the tree. It helps give the tree strength.

Chapter checkup

Write **T** for true or **F** for false in the blanks below.

——— 1. The stem of a plant carries food from the roots to the leaves.

——— 2. Woody stems have chlorophyll and help produce food for the plant.

——— 3. Dicot and monocot plants have herbaceous stems.

——— 4. Stems sometimes bend to allow the leaves to get more sunlight.

——— 5. The outer bark of a tree has phloem cells that carry food from the leaves to the roots.

——— 6. The dicot stem has a thin outside wall called an epidermis.

——— 7. The geranium has a woody stem.

——— 8. Both dicot and monocot stems have vascular bundles.

blade
veins
stalk
epidermis
palisade cells
stomata (and stoma)
chlorophyll

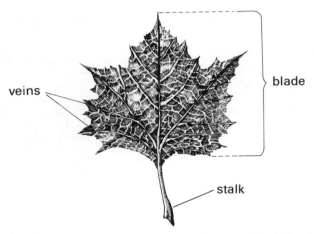

Take a walk in the woods. Look at different leaves. Each leaf has its own special look. Some leaves are quite large. Others are as small as a pinhead. Some have smooth edges. Others have long pointy edges. Try feeling their textures. Some are smooth. Others are rough or even fuzzy. No matter what their size, shape, or texture, they all have an important job. They make food for the plant.

Look at this picture of a maple leaf. The flat part of the leaf is the **blade**. Most food-making cells are in the blade. Throughout the leaf are **veins**. They look like tiny ribs. They help support the leaf. Veins contain xylem and phloem cells like the ones in the root and stem. They do the same job as they do in the root and stem. Xylem cells carry water into the leaf. Phloem cells carry food to the rest of the plant.

At the base of the leaf is the **stalk**. The stalk is like a large vein. It connects the leaf to the stem of the plant.

Try this #19

Materials: leaves from five different plants
1. Carefully examine each leaf.

2. On each leaf locate the blade, the veins, and the stalk.

3. Are any of the leaves missing a part? _____

4. If so, name the leaves and the parts they are missing.

_____

Cells in a leaf

A leaf has several layers. Look at the diagram below.

The outside layer is the **epidermis**. It is the skin of the leaf. It helps protect the leaf from wind, rain, and disease. Under the epidermis are the food-making cells. These are made of two different types of cells. The cells at the top are called **palisade cells**. They are long, thin, and tightly packed. The cells in the lower part are called spongy cells. They are odd-shaped and loosely packed. The veins pass between the food-making cells. The food-making cells absorb water and minerals from the xylem cells in the veins. The food-making cells deposit the food they make into the phloem cells in the veins.

Cross-section of a leaf

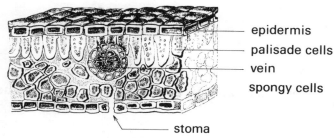

- epidermis
- palisade cells
- vein
- spongy cells
- stoma

There are small openings in the epidermis called **stomata**. (Stomata is the plural. One is called a **stoma**.) Air goes in and out of the stomata. Excess water goes out through the stomata. Around each stoma are two bean-shaped cells. These cells control the size of the opening. They are like balloons. Excess water in the leaf blows them up and enlarges the opening. When there is little water in the leaf, they deflate and close the opening.

 stoma open   stoma closed

Leaves in autumn

The leaves of most plants are green. They are green because they contain **chlorophyll**. Chlorophyll is the green pigment in plants. Chlorophyll also is the substance that takes energy from the sun so the plant can make food.

Pine needles are the leaves of a pine tree. Chlorophyll makes the needles green.

Brian Parker   Tom Stack & Associates

Try this #20

Materials: one board
1. Place the board outside on the lawn in a place where it will not be disturbed.
2. Wait one week.
3. Pick up the board and examine the grass underneath.
4. What changes did you see in the grass?

_____

The grass below the board turned yellow. Light could not reach the grass. Without light, the grass could not produce chlorophyll.

Now you know what happened to the green color, but where did the yellow color come from?

Materials: test tube, water, filter paper, black marker

1. Fill the test tube half full of water.
2. Use the scissors to cut a strip of filter paper about one-half inch wide and three inches long.
3. About half way up the filter paper make a line with the marker.

4. Place the filter paper in the test tube so that the end just touches the water.

5. Wait about 15 minutes. Describe what has happened to the filter paper.

_____

The black dye from the marker is a combination of all colors. The water ran up the filter paper. It flowed through the black dye. It separated the colors.

Materials: beaker, Bunsen burner, matches, filter paper, alcohol, scissors,
ring stand, leaf, two spoons
1. Fill half the beaker with water.
2. Using the Bunsen burner, heat the water to boiling.
3. While waiting for the water to boil, fill half the test tube with alcohol.

4. Use the scissors to cut a strip of filter paper about one-half inch wide and about three inches long.
5. After the water boils, turn off the Bunsen burner and place the test tube of alcohol in the water. (Never boil alcohol over the burner. It could cause an explosion or fire.)
6. Tear the leaf into small pieces and grind it between the spoons.

7. Put the ground leaf into the alcohol.
8. Place the filter paper in the test tube so that just the end touches the alcohol.

9. Wait about 15 minutes. Describe what happened to the filter paper.

___

You probably saw colored layers of green, yellow, and orange. All these colors were in the leaf. In the summer you cannot see the yellow and orange. They are hidden by the green. When the weather gets colder the leaves no longer make chlorophyll. Without the green chlorophyll, the yellow and orange are visible. This is why leaves show more colors in the autumn.

Chapter checkup

Fill in the blanks below with the following words:

stalk   stomata
food   veins
epidermis   blade
chlorophyll

1. The _____ pass between the food-making cells.

2. The outside layer of a leaf is called the _____.

3. Leaves make _____ for the plant.

4. The flat part of a leaf is called the _____.

5. _____ are small openings in the epidermis through which air and water pass.

6. The leaves of most plants are green because they contain

   _____.

7. The _____ connects the leaf to the stem.

64

14. Photosynthesis

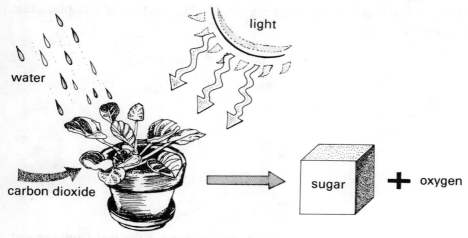

Green plants use energy from sunlight to make food.

All living things need energy. You need energy to work and play. You get this energy from food you eat. Plants don't have to eat food. They make their own food. They make food by a process called **photosynthesis**.

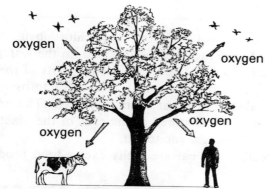

Oxygen is produced through photosynthesis.

How photosynthesis works

Photo means "light." *Synthesis* means "put together." Photosynthesis means "put together with the aid of light." Plants absorb water from the ground. They take in carbon dioxide through the stomata in the leaves. The chlorophyll in the leaves traps light energy from the sun. This energy is used to join the water and carbon dioxide together to form sugar. Sugar becomes food for the plant.

Photosynthesis is simply putting water and carbon dioxide together—with the help of chlorophyll and sunlight—to make sugar.

Materials: green leaf, alcohol, beaker, Bunsen burner, matches, tweezers, ring stand, iodine solution

1. Fill the beaker half full of water.
2. Using the Bunsen burner, heat the water to boiling.
3. While waiting for the water to boil, roll up the leaf and put it in the test tube.
4. Fill the test tube with alcohol until the alcohol completely covers the leaf.
5. After the water begins to boil, turn off the Bunsen burner and place the test tube in the water.

6. After about 15 minutes use the tweezers to remove the leaf from the test tube.
7. Wash the leaf with water.
8. Carefully cover the leaf with the iodine solution. Rinse off the excess iodine.
9. What did you observe happening to the leaf?

Boiling the leaf in alcohol removed the chlorophyll. Without the chlorophyll it is easier to see the effect iodine had on the leaf. Iodine turns brown in the presence of starch. The iodine turned parts of the leaf brown. This showed that there is starch in the leaf. Through photosynthesis plants make sugar. Plants make more sugar than they need. The extra sugar is changed to starch. The starch is stored in the cells of the plant. The iodine showed that starch was in the leaf you tested.

The fruits and vegetables we eat are really excess food produced by the plant.

Photosynthesis produces oxygen

When the water and carbon dioxide combine during photosynthesis, oxygen is given off. All living things, including plants, use this oxygen. Plants use this oxygen to break down sugar and release its energy. Plants produce more oxygen than they use. The excess oxygen is released and used by animals. Without this oxygen, we would not have enough oxygen to breathe. Plants supply almost all the oxygen in the world.

Materials: large jar, green water plant, test tube, several small stones, ring
        stand, clamp, wooden splint, matches, funnel
1. Fill the jar with water.
2. Place 3-4 rocks in the jar.
3. Put the water plants in the jar and cover them with the funnel.

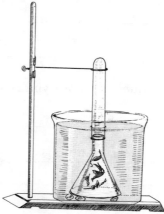

4. Fill the test tube with water and place it upside down over the stem of the funnel.
5. Using the ringstand and clamp, fasten the test tube in place.
6. Place the entire setup near a window where it will receive plenty of sunlight.
7. Wait 24 hours.
8. What happened to the water in the test tube?

_____

9. Where do you think the gas came from?

_____

_____

10. Carefully remove the test tube. Keep your thumb over the opening so that none of the gas inside escapes.
11. Use the matches to light the wooden splint. Let the splint burn for several seconds, then blow it out.
12. Remove your thumb from the mouth of the test tube. Quickly thrust the wooden splint into the test tube.

13. What happened to the wooden splint? _____

In *Try this #24*, the wooden splint burst into flames. This is a test for the presence of oxygen. This oxygen was given off by the plant during photosynthesis.

Chapter checkup

1. Name the four things necessary for photosynthesis.

   1. _____

   2. _____

   3. _____

   4. _____

2. Name the 2 products of photosynthesis.

   1. _____

   2. _____

Science words

reproduction	rhizomes	ovary
spores	bulbs	ovules
cuttings	stamen	sperm
runners	pollen	fertilization
buds	pistil	

15. Reproduction in plants

Living things produce other living things that are like themselves. This is called **reproduction**.

Spores

Simple reproduction is found in the lower forms of plant life. Simple plants, like molds, reproduce by **spores**. Each spore is a tiny cell with a tough cell wall. Simple plants produce millions of spores. The spores are released into the air by the parent plant. These spores may stay alive for years without developing. When a spore lands where conditions are right, it will start to grow.

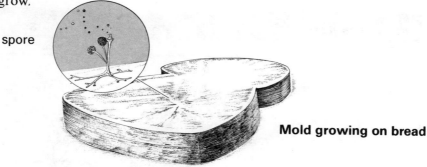

spore

Mold growing on bread

Cuttings

Some higher plants can reproduce from a cutting. A **cutting** is a part of a plant that has been removed. A cutting will grow into a new plant.

Materials: Swedish ivy, glass jar, scissors
1. Use your scissors to cut off the last three inches of one of the stems.
2. Fill the glass jar half full of water and put the end of the stem into the water.
3. Check the water supply every day. Observe what happens to the stem.

You will find the stem has started to grow a new root system. You now have a brand-new Swedish ivy plant.

Runners

Plants like strawberries can reproduce by **runners**. Runners are special stems that grow along the top of the ground. The end of the stem forms a root system. Eventually this root system develops into another strawberry plant. When the new plant is mature the runner dies.

Try this #26

Materials: saxifrage plant in a pot, flower pot with soil
1. Find one of the runners on the saxifrage plant.
2. Place the end of the runner in the second flower pot.
3. Water both the plant itself and the runner every other day. Observe what happens to the runner.

After a while the runner will form roots. When these new roots become established, the runner will die. But you will have a new saxifrage plant.

Buds

Potato plants have stems that grow underground. These stems store food. As the food collects, these stems begin to swell. The swollen parts are the potato. The potato forms **buds** that grow into new plants.

Try this #27

Materials: potato, bowl, knife
1. Cut off a large section of potato. Make sure the part you choose has buds.
2. Put the potato section in the bowl and add enough water to cover about half of it.

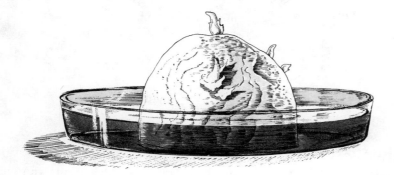

3. Check the water supply every day. Observe what happens to the potato.

The potato is simply part of the stem. It grew into a new plant.

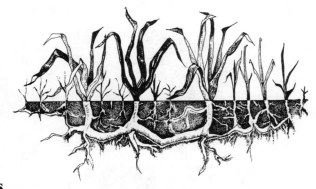

Rhizomes

The grass in your lawn reproduces by **rhizomes**. Rhizomes are underground stems that grow directly into new plants.

Bulbs

Tulips and onions can reproduce from **bulbs**. Bulbs are really part of the stem. The bulb produces new bulbs at its base. Each new bulb grows into a separate plant.

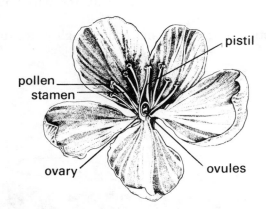

Seed plants

Most plants reproduce from seeds. Even many of the plants that have runners, buds, rhizomes, and bulbs also reproduce by seeds. The seeds are made in the flowers of each plant. Look at the picture above.

Most flowers have both male and female reproductive organs. The male organ is the **stamen**. The stamen produces a powdery substance called **pollen**. The female organ is the **pistil**. At the base of the pistil is a small opening called the **ovary**. Inside the ovary are the **ovules**, or eggs. The ovules are made in the ovary.

Look at the picture of the Easter lily below. It looks different from the flower shown on page 72, yet it has all the same organs. In the Easter lily, find and label the stamen, pollen, pistil, ovary, and ovules.

When the flower matures, the pollen grains loosen from the stamen. Some grains are blown away by the wind. Some become attached to birds and insects. When pollen grains land on another flower of the same kind they attach to the sticky part of its pistil. Reproduction is about to begin.

The pollen grain sprouts a long tube. This tube grows down the pistil to the ovary. A special cell called a **sperm** will move down the tube into the ovary. It attaches to one of the ovules. This process is called **fertilization**. Fertilization is the joining of an ovule (egg) and a sperm cell. The fertilized ovules are the seeds for the next generation.

Bees and other insects are one way pollen is spread.

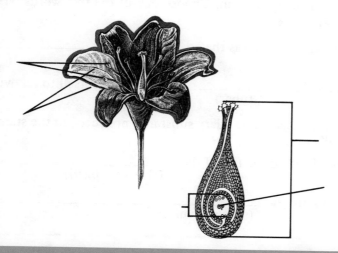

Try this #28

Materials: tulip, single-edged razor blade
1. Take the tulip and remove some of the petals so the inside can be seen.
2. Locate the stamen and the pistil.
3. How many stamens did you find? _____
4. Touch the top of the pistil. Does it feel sticky? _____
5. Remove the pistil.
6. Using your razor blade cut open the base of the pistil. Locate the ovary and the ovules.

7. Check with your teacher to make sure you have properly identified all parts.

Chapter checkup

A. Name seven different ways plants reproduce.

1. _____   5. _____

2. _____   6. _____

3. _____   7. _____

4. _____

B. Fill in the blank in each question. Choose one of the two words written next to the answer blank.

1. The female organ of a plant is the _____
   (pollen, pistil).

2. In fertilization a _____ cell attaches itself to an ovule.
   (sperm, tuber)

3. _____ is the powdery substance produced in the stamen.
   (Pollen, Ovule)

4. Pollen attaches itself to the top of the _____ .
   (rhizome, pistil)

5. Inside the ovary are the _____ .
   (sperm, ovules)

Unit review

- Plants and animals depend on each other. Plants produce all the oxygen that animals need to live. Animals breathe the oxygen and produce the carbon dioxide that plants need.
- Most plants have roots, stems, and leaves. Some plants also have flowers and fruits.
- There are two types of roots—fibrous and taproots.
- Roots have three main parts—epidermis, cortex, and central core.
- Xylem cells throughout the plant carry water and dissolved minerals up the root and stem to the leaves.
- Phloem cells carry food made in the leaves down through the stem to be stored in the root.
- Herbaceous stems are soft and green. Monocot and dicot plants have herbaceous stems.
- Trees have woody stems.
- Plants make food in their leaves.
- Air and water pass in and out of the leaves through small openings called stomata.
- Chlorophyll is the green pigment in plants.
- Plants use chlorophyll to trap energy from the sun for photosynthesis.
- Photosynthesis is the process by which the plant makes its own sugar and oxygen. The plant needs water and carbon dioxide and sunlight and chlorophyll for photosynthesis.
- Excess sugar is stored as starch in the plant.
- Plants reproduce by a variety of means, but most reproduce by seeds.
- The male part of the plant is the stamen. It produces pollen.
- The female part of the plant is the pistil. Inside is the ovary. It produces ovules.
- A sperm cell fertilizes the female ovule of the plant to produce seeds for new plants.

Unit checkup

Reproduction

1. In a flower, the male reproductive organ is called the

   _____.

2. The powdery substance produced in the stamen is _____.

3. The female reproductive organ in a flower is the _____.

4. Inside the ovary are the eggs, or _____.

5. When a sperm cell attaches to one of the ovules, _____ has taken place.

Plant parts

Label the parts of the diagram below. Use these terms: stamen, pistil, ovary, root, leaves, stem.

True or false

_____ 1. Plants can't grow in a desert.

_____ 2. Plants make the carbon dioxide that animals need to survive.

_____ 3. Plant roots take in water and dissolved minerals.

_____ 4. You can eat the taproots of some plants.

_____ 5. Xylem cells store oxygen.

_____ 6. The veins in a leaf contain the food-making cells.

_____ 7. The root cortex covers and protects the outside of the root.

_____ 8. Chlorophyll is one of the substances necessary for photosynthesis.

_____ 9. The vascular bundles inside the plant stem contain xylem and phloem cells.

_____ 10. The stalk of a leaf connects the leaf to the stem.

_____ 11. Stomata are the male reproductive parts of a flower.

_____ 12. Leaves turn color in the autumn because they no longer make chlorophyll.

_____ 13. Photosynthesis produces sugar and oxygen.

_____ 14. The female organ of the plant is the pollen.

_____ 15. The female organ produces rhizomes in its ovary.

Unit 4.
Humans and other animals

Just like plants, humans and other animals have basic needs. These needs must be met if the animal is to survive. Animals must be able to obtain food. This food must have a way to be digested. Once digested, it must be carried to every cell in the animal. Animals must have oxygen. Just like food, oxygen must be carried to every cell. Once the food and oxygen have been used by the cells, there must be a way to remove the waste products. Animals also must be able to react to their surroundings. They must be able to move toward things they need and away from things that may harm them. Finally, animals must be able to reproduce so the species can survive. So get yourself ready for a peek into the life functions of animals, especially humans!

zoologist
digestion
circulation
respiration
excretion
hormones
endocrine
nerves
muscular
skeletal
reproductive system

16. Animals

Dwight R. Kuhn

How long do you think it would take to write the name of every animal that exists today? If you wrote one name every second, you would have to write day and night for more than twelve days to finish the list. There are more than one million different types of animals in the world.

In chapter nine, you learned that botanists are scientists who study plants. Scientists who study animals are called **zoologists**. In many ways, their jobs are similar. Plants and animals have a lot of things in common.

Plants and animals are very much alike

In the last unit, you studied plants. Plants perform certain life functions. Animals perform many of the same life functions.

Both plants and animals need food to survive. Plants are able to make their own food. Animals cannot. Animals use food made by plants. Animals must change the food into a form they can use. This process of changing food into a usable form is called **digestion**. This is done in the digestive system.

Plants transport water and minerals from their roots to their leaves. Animals also have a system of transporting food and water. In animals, blood transports the food and water. This is called **circulation**. This is done in the circulatory system.

Both plants and animals need oxygen to survive. Animals use oxygen to release energy from food. This is called **respiration**. In animals, this is done in the respiratory system. Breathing is part of respiration.

In plants, oxygen is a waste product of photosynthesis. Carbon dioxide is a waste product of respiration in animals. The elimination of carbon dioxide and other wastes is called **excretion**. This is done in the excretory system.

Plants produce chemicals that help them change. Animals also produce chemicals. These chemicals are called **hormones**. Hormones help the animal's body to change. Hormones are made in the **endocrine** system.

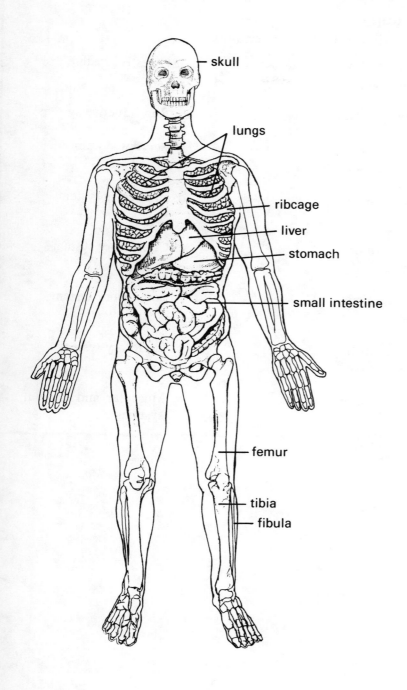

skull

lungs

ribcage

liver

stomach

small intestine

femur

tibia

fibula

Both plants and animals respond to changes around them. Plants generally respond very slowly to changes. Animals respond much faster because they have **nerves** that detect changes around them. Nerves are a part of the nervous system. Another part of the nervous system is the brain. Plants do not have brains.

Many plants have stems to hold them up. Many animals have bones and soft tissue to give support, to hold them up, and to allow them to move. These are called the **muscular** and **skeletal** systems.

Both plants and animals create offspring like themselves. Plants have many ways of reproducing. Animals reproduce sexually. Male and female animals have **reproductive systems** for creating new life.

Chapter checkup

Match the system with the function it performs.

1. Transports food and water throughout the body.

    a. Digestive system

2. Rids the body of wastes.

    b. Reproductive system

3. Enables a species to create more of its own kind.

    c. Nervous system

4. Converts food into a usable form.

    d. Endocrine system

5. Takes in oxygen to be used by the body.

    e. Circulatory system

6. Detects changes outside the body.

    f. Respiratory system

7. Produces chemicals that help the body to change

    g. Excretory system

8. Gives support to the animal

    h. Muscular and skeletal systems

cerebrum
cerebellum
medulla
spinal cord
neurons
dendrites
axons
synapse
sensory neuron
motor neuron
reflex action

17. The nervous system

A fly lands on your arm. You try to swat it. You miss. The fly quickly reacts and is gone.

The fly has a nervous system. Its nervous system tells it to move. The fly saw your hand coming. This information was sent along the nerve cells to its brain. The brain sent back information along another line of nerve cells. The information told the wings to move. The fly moved before you could swat it. All this happened in an instant. The nervous system works very quickly.

R. Calentine

Nerves in the earthworm sense light but not sound.

The nervous system of lower animals

Many lower animals have very simple nervous systems. The sponge, for example, has special nerve cells on the outside of its body. These cells send messages to the cells on the inside. The cells work together to help the sponge gather food.

In the jellyfish, all the nerve cells are connected. If one part of the jellyfish is touched, its entire body reacts.

Worms have a very well-developed nervous system. Worms have a "brain." They have a nerve cord that is connected to the "brain." This cord runs the entire length of the worm's body. Information is sent to and from the "brain."

Tom Stack & Associates

Nerve cells in the sponge are connected throughout its body.

Eagles have a keen sense of sight.

The nervous system of higher animals

Vertebrates have the most highly developed nervous systems. This is because they have highly developed brains. Different parts of the brain receive different signals. In a fish, the part of the brain designed for smell is large. Fish have a keen sense of smell. In an eagle, the part of the brain that controls sight is very large. This allows eagles to spot prey at great distances. Most mammals are capable of learning. This part of the brain is very large in mammals.

Let's look at the nervous system of humans. Like that of other vertebrates, it is made up of the brain, the spinal cord, and nerves.

Human brain

Your brain is the control center of your body. It receives information. It sends out instructions for responses. It tells your heart to beat and your lungs to breathe. It tells your muscles to make your hand or leg move. It tells your mouth to open or close. Your brain allows you to think, see, hear, and smell.

The top of your brain is the **cerebrum**. The cerebrum controls learning. The cerebrum has two sides. The right side of the cerebrum controls the left side of your body. The left side of the cerebrum controls the right side of your body. The left side of the cerebrum is the center of speech and logical thinking. The right side is the center of creativity and feelings. The left side thinks more with words and looks at details. The right side thinks in pictures and looks at the whole, rather than at details.

The part of the brain that controls smell is well developed in fish.

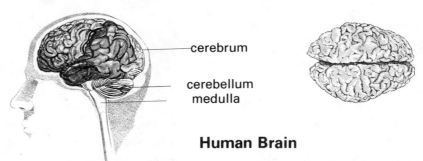

cerebrum

cerebellum
medulla

Human Brain

Below your cerebrum is your **cerebellum**. It is much smaller than your cerebrum. It is at the base of your cerebrum. Your cerebellum controls your muscles and movement of your body after you have "thought" about moving.

Your **medulla** is also at the base of your cerebrum. Your medulla controls your internal organs like your heart, lungs, stomach, and so on. It is controlling the muscles that you do not consciously think about moving.

Spinal cord

Your **spinal cord** comes out of the base of your brain. Your spinal cord is a large bundle of nerves. Information to and from your brain travels through your spinal cord. Nerves to all parts of your body lead out of your spinal cord. Nerves from all parts of your body come into your spinal cord.

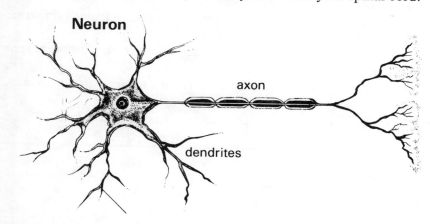

Neuron

axon

dendrites

Neurons

Your nervous system is made up of millions of **neurons**. Neurons are nerve cells. These special cells receive information. They transfer this information to other parts of your body.

Neurons do not look like most other cells. One end is round with short branches coming out of it. These branches are called **dendrites**. The other end is long and thin. This long thin part is called the **axon**.

Neurons work together to send information. They line up end to end. The information enters the cells through the dendrites. It is transferred out of the cell through the axons. This assures that information will travel in only one direction.

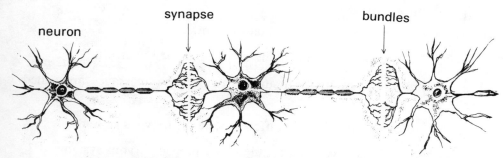

synapse   bundles

neuron

Messages jump across the synapse to the next neuron.

Neurons do not actually touch each other. There is a very small space between the dendrite of one cell and the axon of the next. That space is called a **synapse**. Messages jump across the synapse to the next nerve cell.

These neurons send information up through your spinal cord to your brain. Your brain makes a decision. The decision from your brain is sent through your spinal cord to other parts of your body.

The nerves that carry messages to your brain are called **sensory neurons**. If you rub your hand across a piece of sandpaper, sensory neurons carry the message of "rough" to your brain. Your brain sends messages back to your body through **motor neurons**. Motor neurons control your muscles. Your brain may tell your muscles not to rub the sandpaper again!

Materials: beaker of cold water, eyedropper
1. Choose a partner.
2. Using the eyedropper, place one drop of water on your partner's forearm.
3. Write down whether he or she felt the cold.
4. Repeat this four more times. Each time drop the water on a different spot on the forearm. Record the answer.
5. Now drop water on your partner's neck. Try it in four different places. Record whether he or she felt the cold.
6. Which has more cold sensing nerves, the forearm or the neck?

Reflex actions

Choose a partner and wave your hand about two inches in front of his or her eyes. Did your partner blink?

The blinking was a **reflex action**. A reflex action is movement without thinking. The brain in not involved. If the nerve impulses had to go all the way to the brain, the reaction might be too slow.

When you touch a hot frying pan, a nerve impulse goes directly to your spinal cord. Your spinal cord stimulates a motor nerve. The motor nerve sends an impulse back to your hand. You pull your hand away without even thinking about it.

Anthony Potter

Pulling away from heat is a reflex action. Reflex actions are quick because the nerve impulse does not have to travel all the way to the brain.

Chapter checkup

Write **T** for true or **F** for false in the blanks below.

_____ 1. The nervous system is made up of three long neurons.

_____ 2. The branches in the neurons that receive messages are called axons.

_____ 3. Different animals have different types of nervous systems.

_____ 4. Vertebrates have the most highly developed nervous systems.

_____ 5. In vertebrates, different parts of the brain control different reactions.

_____ 6. The cerebellum controls thinking, learning, and talking.

_____ 8. Sensory neurons carry information from the brain to various parts of the body.

_____ 9. Motor neurons control the muscles.

_____ 10. Solving a difficult math problem is a reflex action.

_____ 11. The brain is not involved in a reflex action.

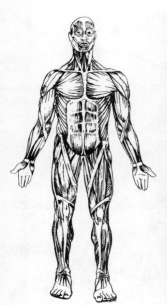

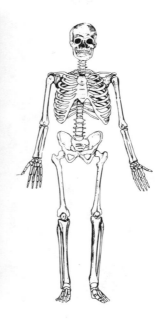

skull
spinal column
rib cage
pelvis
joint
cartilage
tendon
ligament
contract
flexing

18. Muscular and skeletal systems

There is an amazing structure or framework for most animals' bodies. They are the muscular and skeletal systems. These systems support and protect the organs of the body. They keep the body from being a shapeless blob! The skeleton is made up of the bones. Muscles are soft tissues.

Skeleton

Some animals have a skeleton outside their bodies. Lobsters and snails have outside skeletons. Humans and most other animals have inside skeletons. This inside skeleton is made up of bones.

Your body has more than 200 bones. Some of the main parts are the skull, spinal column, rib cage, and pelvis.

Your **skull** protects your head. Your **spinal column** supports your head. It also protects your spinal cord, which is the main nerve. Other major bone structures, like your shoulder structure, ribs, and pelvis, attach to your spinal column.

Your **rib cage** expands when you breathe. It protects your heart, lungs, and other organs. Your arms attach to your shoulder structure. Your **pelvis** is the hip area of your body. Your pelvis supports the weight of your body. Your legs are connected to your pelvis.

Some bones move and some do not. The bones that move most easily join with other bones at **joints**. Your elbow is a joint; so is your knee. There are many joints in your hands and feet. **Cartilage** is a connective tissue. It reduces friction between the two bones of the joint. It allows your bones to move easily without rubbing against each other.

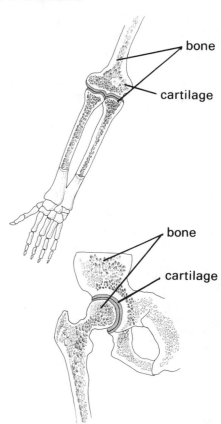

Cartilage prevents bones from rubbing against each other.

Human Skeleton

- skull
- ribcage
- spinal column
- pelvis

Skeletal Muscles

- trapezius
- pectoralis major
- biceps
- rectus abdominus
- sartorius

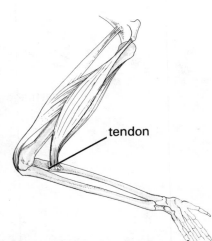

- tendon

Tendons connect some muscles to bone

Muscle system

There are two kinds of muscle in your body—skeletal muscles and internal muscles. The internal muscles are organs like your heart, stomach, and intestines. (We will look at these in other chapters.) These internal muscles or organs move *involuntarily*. That is, you don't have to tell your heart to beat or your stomach to digest food. The skeletal muscles are voluntary muscles. You must order your arms or legs or hands to move.

Your skeletal muscle system moves your body. Other connective tissues besides the cartilage help your muscles and bones do their jobs. The **tendons** connect some muscles to the bones. **Ligaments** connect bones to each other.

Skeletal muscles also might be called connective tissue. Muscles are flexible. They can stretch like rubber. Most muscles connect one bone to another bone. Some muscles connect directly with the bone. Others are connected to the bone by a tendon.

Muscles are usually in pairs. Suppose you want to bend your arm. A message from your brain travels down the nervous system to the muscles in your arm. The message tells one muscle to **contract**. The muscle becomes shorter and thicker. When you contract a muscle, you are **flexing** it. This brings the two bones closer together. Contracting one muscle causes the other muscle of the pair to extend or stretch a little. This flexing of one muscle and the stretching of the other muscle causes your arm to bend.

There are muscles in your stomach, chest, shoulders, back, arms, legs, face, and neck. There are muscles throughout your entire body. Exercising muscles keeps them strong and healthy. Look at the parts of your body you use most often. Are you lefthanded? Look at your left arm. Is it stronger than your right arm? The muscles are used more, so they are in better shape.

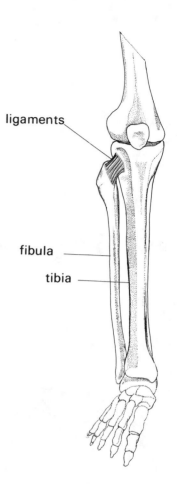

ligaments

fibula

tibia

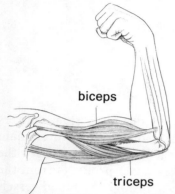

biceps

triceps

The bicep and tricep muscles in your arm work as a pair. When the bicep contracts, your arm bends. When your tricep contracts, your arm straightens.

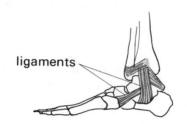

ligaments

Chapter checkup

skull	cartilage
spinal column	tendons
pelvis	ligaments

Fill in the blanks.

1. The part of your skeleton that protects your head is your _____.

2. The _____ supports the head; many other important parts of the skeleton are attached to it, too.

3. The _____ connect bone to bone.

4. The _____ connect some muscles to the bones.

5. The _____ reduces friction between two bones at the joint.

6. The _____ supports the weight of your body.

Science words

arteries
aorta
capillaries
veins
atrium
ventricle
plasma
platelets

19. The circulatory system

All cells need food, water, and oxygen. The circulatory system delivers these things to the cells. The circulatory system is made up of three main parts: the blood, the blood vessels, and the heart. The blood carries the food, water, and oxygen to the cells. It also takes waste products from the cells. Blood vessels are like pipes. The blood flows through the blood vessels on its way to and from the cells. The heart works like a pump to push the blood through the blood vessels.

The human circulatory system is made up of the heart, the blood, and all of the blood vessels.

Open and closed systems

There are two types of circulatory systems. Most lower animals have open systems. Most higher animals, including humans, have closed systems.

In an open system, the blood does not stay in the blood vessels at all times. It flows directly over the cells. In doing so it drops off the food, water, and oxygen. It picks up wastes. It then returns to the blood vessels. The crab and the crayfish have open systems.

In a closed system, the blood never leaves the blood vessels. The food, water, and oxygen pass through the thin vessel walls to the cells. The wastes pass back through the vessel walls to the blood.

The blood vessels

Blood vessels are like pipes that carry blood throughout the body. There are three different types of blood vessels. **Arteries** carry blood *away* from the heart. The largest artery in the body is the **aorta**. It comes out of the left ventricle carrying blood with fresh oxygen. The aorta then breaks off into other large arteries. Those large ateries break into smaller arteries. The smaller arteries carry blood to all parts of the body. Arteries break into still smaller vessels called **capillaries**. Capillaries are microscopic blood vessels. Here the blood passes food, water, and oxygen to the cells. The blood picks up carbon dioxide and other wastes through the capillaries.

Pathway of Blood

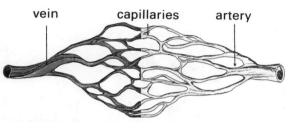

Capillaries carry blood from artery to vein.

The capillaries empty into **veins**. Veins are the blood vessels that carry the blood back to the heart. The capillaries empty into very small veins. The small veins branch into larger and larger veins. Finally, the larger veins branch into the main veins that go into the heart.

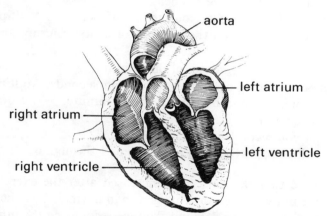

Parts of the Heart

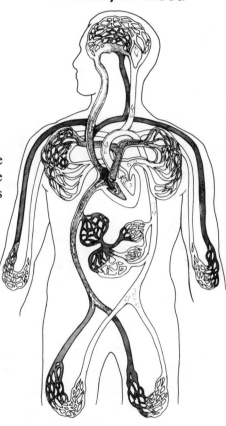

■ **Blood high in carbon dioxide**

□ **Blood high in oxygen**

Heart

The heart works like a pump. It pumps the blood throughout the circulatory system.

Your heart is about the size of a large fist. It is located under your breastbone, between your lungs. Your heart has four chambers. The two upper chambers are called **atriums**. There is a left atrium and a right atrium. The two lower chambers are called **ventricles**. There is a left ventricle and a right ventricle.

The left atrium receives blood with fresh oxygen from your lungs. The left atrium pumps this blood to the left ventricle. The left ventricle pumps this blood out to your entire body (except the lungs). The blood travels through your body and then back to your heart through the right atrium. The right atrium pumps the blood down to the right ventricle. The right ventricle then pumps the blood up to your lungs for more oxygen. The blood just keeps on flowing through your heart and body in the same endless circle.

Materials: watch with second hand, paper, pencil

1. Choose a partner. Put the tips of your two middle fingers on the inside of your partner's wrist.

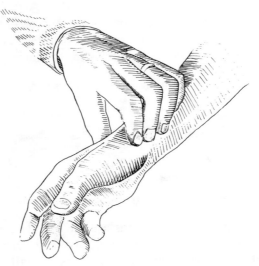

2. Press firmly until you feel a pulse or "beat."
3. Count the number of pulse beats you feel for 15 seconds. Multiply that number by 4. That will tell you the number of beats in a minute. Write down the number of beats per minute. _____
4. Now have your partner run in place for one minute.
5. Count the number of pulse beats you feel for 15 seconds. Multiply that number by 4 to find the number of beats per minute. Write down the number of beats per minute. _____
6. In which case did you record the higher pulse rate? _____
7. Switch places with your partner and repeat steps 1 through 6.

   You probably found that the pulse rate increased after the exercise. The exercise caused your body to use more oxygen than normal. Your heart had to beat faster to keep up with the demand. You were counting the number of times per minute the heart was beating. The heart beats each time the ventricles pump blood out.

Circulation

Circulation in most higher animals follows the same general pattern. The heart pumps blood to the lungs. The blood that has circulated through the body leaves the waste product, carbon dioxide, for the lungs to get rid of. The blood picks up oxygen in the lungs. The blood picks up food for the cells in the small intestine. Blood that passes through the digestive organs passes through the liver and kidneys before it is returned to the heart. The liver and kidneys remove harmful materials that the blood might have picked up.

Circulatory System of a Frog

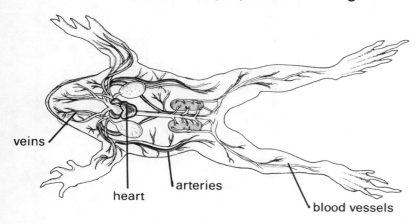

veins

heart

arteries

blood vessels

Things to do

Look at the word *circulatory*. In it you can almost see the word circle. Why do you think this is called the circulatory system?

_____

Blood

Human blood has four different parts. **Plasma** is the liquid part of your blood. It is made mostly of water. It carries the food to the cells. It carries waste materials away from the cells. The other three blood parts are carried in the plasma.

Red blood cells carry oxygen to body cells. They carry waste carbon dioxide away from the cells. Blood looks red because of the many red blood cells in it. Oxygen makes the red blood cells red.

White blood cells fight diseases that try to enter your body. They also fight infection.

Platelets cause blood to clot. This is important when you cut yourself. Platelets stop the blood from flowing out. Then the wound can heal.

Chapter checkup

Across
1. The blood, heart, and blood vessels make up the _____ system.

3. _____ blood cells carry oxygen.

4. The part of the blood that carries food to the cells is the _____.

6. In a human heart, the left _____ pumps blood to the body.

8. _____ are microscopic blood vessels.

Down
2. In the human heart, the right _____ receives blood that has circulated through the body.

5. _____ carry the blood away from the heart.

7. _____ blood cells fight disease.

9. _____ cause the blood to clot.

10. _____ carry blood back to the heart.

20. The respiratory system

The air around us has a number of different gases in it. Most of it is a gas called nitrogen. Nitrogen makes up about four-fifths of the air. About one-fifth of the air is oxygen. There are also small amounts of carbon dioxide and other gases.

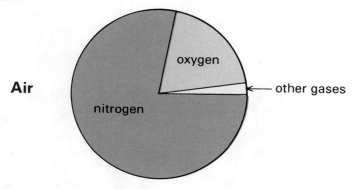

Try this #31

Materials: glass jar, straw, limewater
1. Fill the jar with limewater.
2. Put a straw into the limewater.
3. Take a very deep breath of air.
4. Blow the air through the straw into the jar of limewater. Blow into the straw several times, if necessary, until you see something happen to the limewater.
5. What did you observe happening to the limewater? _____

_____

What do you think caused it? _____

_____

When you blew through the straw, the limewater turned milky. Something in the air you exhaled caused the limewater to change. The gas that caused this was carbon dioxide. Your body takes in oxygen and gives off carbon dioxide. Carbon dioxide is a waste product of breathing.

Most fish breathe through gills located on the side of the head. Gills remove oxygen from the water.

Respiration

All living things need oxygen to live. Taking in oxygen and giving off waste products, like carbon dioxide, is called **respiration**. Every living cell needs oxygen to burn food to make energy. Think of a fire. It needs air so that the wood can burn and make heat, which is energy. An animal needs oxygen so that it can burn its food and make energy for living.

Animals have different ways of taking in oxygen and getting rid of carbon dioxide. Single-cell creatures take oxygen in through the cell membrane. Fish have gills that take oxygen out of the water. Earthworms take oxygen in through their skin. Insects have special holes to take in air. Most land vertebrates, including people, have **lungs** for taking in oxygen.

Many insects get oxygen through tiny holes on the sides of their bodies.

Lungs

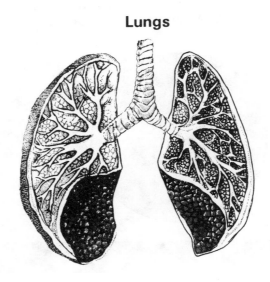

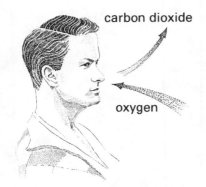

carbon dioxide

oxygen

Respiration involves taking in oxygen and giving off carbon dioxide and other waste products.

Respiration in humans

When you breathe, you take air into your lungs and let air out of your lungs. When you take air in, you **inhale**. When you let air out, you **exhale**.

Air enters your body through your nose or your mouth. When you breathe through your nose, the hairs in your nose act as filters. They remove dust from the air. The air travels down your windpipe into your lungs. There are other filters along the way to take out even smaller bits of dust in the air. Finally, the air enters your lungs. Your lungs are very soft and spongy. That's because inside the lungs there are millions of air sacs. The air sacs are surrounded by capillaries (very tiny blood vessels). Oxygen from the air in the lungs passes into the blood stream through the capillaries. The blood-stream takes the oxygen to all the cells in your body. Carbon dioxide waste passes out of the blood stream through the capillaries. It travels to your lungs. Then the carbon dioxide is exhaled with the air in your lungs.

The diaphragm

The power for breathing comes from your **diaphragm**. Your diaphragm is a set of muscles just below your chest. These muscles contract and pull down making your chest cavity larger. Air rushes in through your nose and mouth to fill the extra space. When the muscles in your diaphragm relax, the chest cavity gets smaller. Air is pushed out. Your diaphragm is constantly contracting and relaxing. Breathing is the contracting and relaxing of your diaphragm. Your diaphragm makes you inhale and exhale.

This next *Try this* helps to show how the diaphragm works.

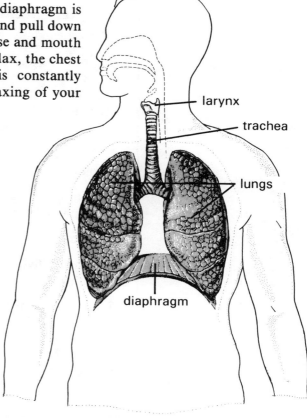

The balloons inflate when the rubber is pulled. Your lungs inflate when your diaphragm contracts.

Try this #32

Materials: Gallon jug with bottom removed, rubber sheeting, Y-shaped tube, 2 balloons, 1-holed cork

1. Attach the balloons to the two smaller ends of the Y-shaped tube.
2. Insert the cork into the mouth of the jug.
3. Insert the end of the Y-shaped tube into the cork.
4. Place the rubber sheeting across the open end of the jug and secure it with a rubber band.
5. Pinch the middle of the rubber sheet and pull down.
   What happens to the balloons? _____

   _____

6. Release the rubber sheeting. What happens? _____

   _____

The rubber sheeting is like your diaphragm. When your diaphragm muscles contract, they pull down (just like when you pulled down on the rubber sheeting). This fills your lungs with air (just as the balloon filled with air). When you released the rubber sheeting, the air left the balloons.

95

Chapter checkup

Fill in the blanks with the following words:

respiration   milky
nitrogen   breathe
diaphragm   oxygen
carbon dioxide   lungs

1. To _____ means to draw air into the lungs then force it out.

2. Your _____ is a set of muscles just below your chest that help you breathe.

3. About four-fifths of the air around us is the gas _____.

4. When a living thing takes in oxygen and gives off carbon dioxide, this is called _____.

5. Your body needs the gas _____ to survive.

6. Your body exchanges oxygen for _____ when you breathe.

7. The exchange of oxygen and carbon dioxide takes place in your _____.

digestion
esophagus
stomach
enzymes
small intestine
saliva
peristalsis
hydrochloric acid
bile

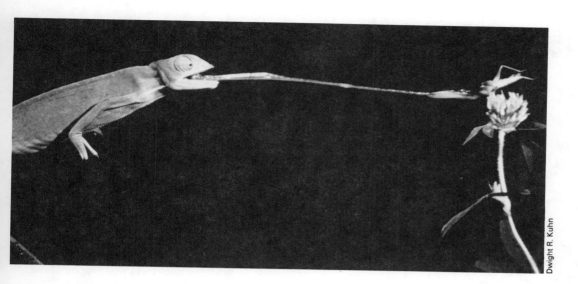

Dwight R. Kuhn

21. The digestive system

All animals need energy to survive. They need energy to carry on life's processes. They get this energy from the food they eat. Food is fuel for making energy. In the last chapter, we said that the body needs oxygen in order to burn that food or fuel. But most food is not in the right form when eaten. It must be broken down. **Digestion** is a process that changes food into a form that can be burned by the cells.

The digestive system

Different kinds of animals have different kinds of digestive systems. Most systems, however, follow the same general path.

Food enters the animal through the mouth. Some animals have teeth. They use their teeth to grind up the food. Grinding the food is the first step in digestion. The mouth contains a liquid. The liquid aids in digestion.

From the mouth, the food passes into the **esophagus**. The esophagus is a tube that leads to the digestive organs. In most vertebrates, the esophagus leads to the **stomach**. The stomach is like a sack. It can contain a large volume of liquid. In the liquid are **enzymes**. Enzymes are chemicals that help break down the food more.

From the stomach, the food goes into the **small intestine**. The small intestine in humans is a tube that is about 20 feet long. But it is folded around so that it fits into a small space. The last steps of digestion happen in the small intestine. More enzymes break down the food even further. Then the food is ready to be used by the body. The digested food passes through the walls of the small intestine. It is picked up by the blood. The blood carries it to the cells of the body.

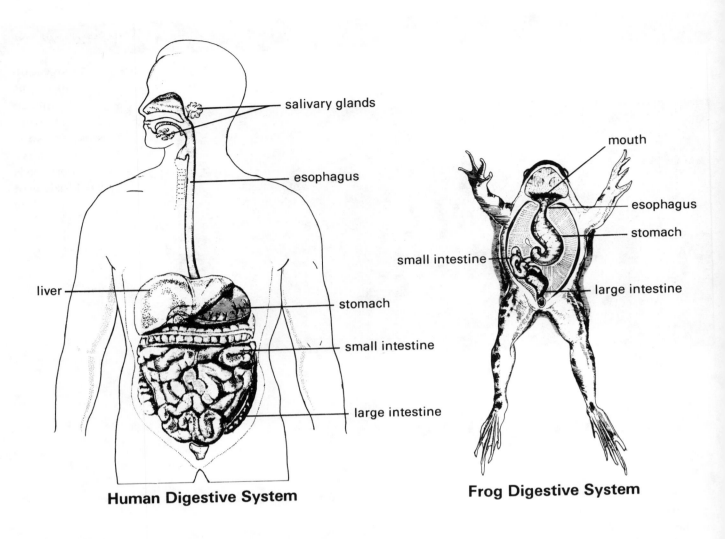

Human Digestive System

Frog Digestive System

Your digestive system vs. a ham sandwich

Sitting before you is a ham sandwich. After you eat and digest it, it will give you energy.

You bite into it and chew. Your teeth tear the piece of sandwich apart. The piece of sandwich becomes wet from the liquid in your mouth. This liquid is called **saliva**. There are enzymes in your saliva. These enzymes are starch-digesting enzymes. They begin to chemically break down the bread. The ham and any fat in the ham are not affected. The enzymes in your saliva affect only the bread because the bread is starch.

You swallow. Swallowing forces the ham sandwich down into your esophagus. **Peristalsis** will keep the food moving down your esophagus and through the rest of your digestive system. Peristalsis is a wave of contractions in the muscles of the digestive system.

Once the food is in your stomach, more enzymes continue to break down the bread. Other enzymes begin to break down the ham. To help the enzymes, your stomach produces a special acid. This is **hydrochloric acid**. Hydrochloric acid works with your enzymes to speed up digestion. The enzymes and the acid work together to break down the ham and bread. The fat remains untouched. There is nothing in your stomach to break down fat.

After three hours, the ham sandwich moves from your stomach to your small intestine. Your small intestine senses the undigested fat. Your liver comes to the rescue. Your liver makes a special chemical called **bile**. Bile breaks down fat. Your small intestine produces still more enzymes. These enzymes finish the digestive process on the bread, ham, and even the fat. The ham sandwich is now fully broken down. It no longer even looks like food; it is microscopic.

The digested food passes through the walls of your small intestine. It is picked up by your blood. The blood will carry it throughout your body. (Parts of the food that can't be used or broken down are waste. We will see what happens to waste in the next chapter.)

When digested, the gazelle will provide fuel for this cheetah and her cub.

Try this #33

Materials: two small glasses, iodine, cracker
1. Place about three tablespoons of water into each glass.
2. Break off two small pieces of cracker, crush them with your fingers and place them in each of the two glasses.
3. Add one drop of iodine to one of the two glasses.
4. What color did the water turn? _____
5. Work some saliva up in your mouth. Spit it into the second glass.
6. Add one drop of iodine to the second glass.
7. What color did the water turn? _____
8. Look at the glasses again in 20 minutes. Notice if there was any color change? If so, in which glass? _____

Iodine is used to test for starch. Iodine turns almost black in the presence of starch. Crackers have starch in them. The iodine in both glasses should have turned black. After about 20 minutes, you should have seen a change in the glass with the saliva in it. The saliva in your mouth contains enzymes. These enzymes changed some of the starch to sugar. If you used enough saliva, all of the starch will have changed to sugar. There will no longer be a black color. Changing starch to sugar is part of the process of digestion. As you can see, digestion starts in your mouth.

Materials: two six-inch sections of sausage skin, string, corn starch, honey, iodine, Benedict's solution, two glasses

 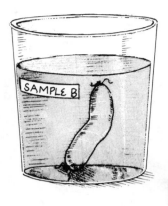

1. Tie one end of each sausage skin closed with a piece of string.
2. In one sausage skin place one teaspoon of cornstarch and two teaspoons of water.
3. Tie the end closed with a piece of string.
4. Place the skin with the cornstarch in a glass of water and mark this sample A.
5. In the other skin, place two teaspoons of honey and two teaspoons of water.
6. Tie the end closed with a piece of string.
7. Place the skin with the honey in a glass of water and mark this sample B.
8. Let these samples sit for one day.
9. Take the sausage skins out of the water and throw them away.
10. Add one drop of iodine to the glass of water marked sample A. Describe the results. _____

    _____

11. Add three drops of Benedict's solution to the glass of water marked sample B. Describe the results. _____

    _____

Remember iodine is a test for starch. It turns black in the presence of starch. The iodine in sample A did not turn black. The corn starch was not able to go through the sausage skin into the water.

Benedict's solution is a test for sugar. It turns red in the presence of sugar. The Benedict's solution turned red. Honey is a sugar. The sugar was able to go through the sausage skin.

The sausage skin came from the intestine of an animal. It is very much like the intestine in your body. It did not allow starch to go through it. Your body cannot use starch. Your body can use sugar. Your intestine holds the starch until it is changed to sugar.

Chapter checkup

Use the words listed below to fill in the blanks:

energy   liver
hydrochloric acid   Benedict's solution
iodine   esophagus
small intestine   sugar
saliva   mouth

1. The _____ is a long tube that connects the mouth to the stomach.

2. Digested food passes through the walls of the _____.

3. Your stomach produces _____.

4. Digestion starts in the _____.

5. _____ turns black in the presence of starch.

6. Changing starch to _____ is part of the digestive process.

7. _____ turns red in the presence of sugar.

8. The _____ in your mouth contains enzymes and helps begin the digestive process.

9. Animals need _____ to carry on life's processes.

10. Your _____ makes bile to break down fat.

excretion
colon
anus
urea
uric acid
kidneys
urine
bladder
pores

22. The excretory system

An animal's body is like a furnace. A furnace must have fuel to operate. Furnaces often burn wood, coal, or oil. Animals need fuel, too. Animals use food for fuel. A furnace leaves wastes. Ashes are the waste products. If the furnace is to operate properly these waste products must be removed. As an animal's body burns food, it also produces waste products. These waste products must be removed. If they are not removed, the animal can become sick and die. The removal of these waste products is called **excretion**. Excretion takes place in the excretory system.

Types of waste

Excretion is ridding of wastes from the body. There are three types of waste products:

Solid wastes—Some parts of food cannot be digested. Undigested food forms solid wastes.

Liquid wastes—Liquid waste is water with dissolved materials in it. The dissolved materials are waste products of the cells.

Gaseous wastes—Carbon dioxide is the waste product of respiration (see chapter 20).

Excretory systems of different animals

Simple animals tend to have simple excretory systems. The more complex the animal, the more complex its excretory system is. The single-celled ameba expels all wastes through its cell membrane. Flatworms are more complex. Flatworms have many openings in their bodies. Wastes collect and are expelled through these openings. Roundworms are still more complex. Roundworms have two separate openings for liquid and solid wastes. Vertebrates have the most complex systems. We will look at the human system because most vertebrates have an excretory system similar to that of humans.

Excretory Organs

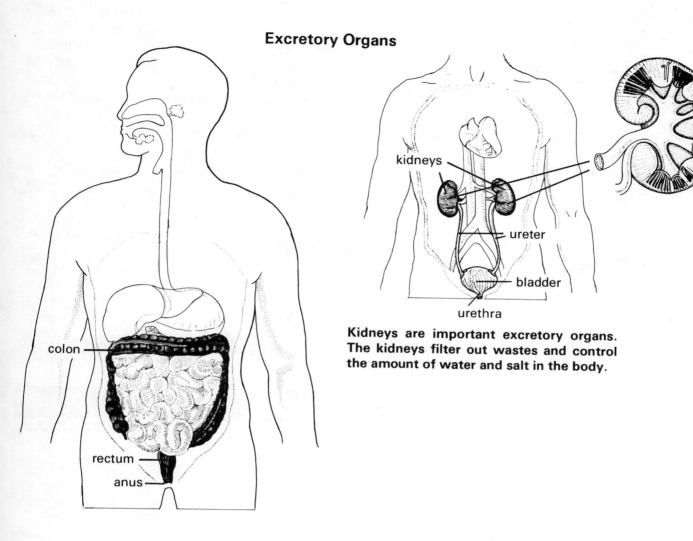

Kidneys are important excretory organs. The kidneys filter out wastes and control the amount of water and salt in the body.

The human excretory system

Elimination of solid wastes. The last stage of digestion takes place in your small intestine. All digested food passes from your small intestine into your bloodstream. But there are parts of the food that cannot be digested. These undigested parts leave your small intestine and enter your **colon**. The colon is commonly called the large intestine. The undigested parts still have a large amount of water in them. Your body needs this water. In your colon, much of this water is absorbed into your blood stream. When enough solid waste collects, it moves through your colon to your **rectum** and then out of your body through an opening called the **anus**.

Elimination of liquid wastes. Food that was digested in your small intestine is carried to the cells by the bloodstream. The cells use the food. The cells produce waste from using the food. Much of this waste is called **urea**. Another waste is **uric acid**.

Urea and uric acid are picked up by the blood. These wastes are like poison. They must be removed from your body quickly. Your **kidneys** perform the very important job of filtering the blood to take out these poisons. The kidneys are bean-shaped organs. Most people have two kidneys. They are located on either side of your spine just below your waist.

Perspiration is one way your body excretes wastes.

As the blood flows through your kidneys, urea and uric acid are filtered out. Your kidneys hold back a small amount of water from the blood. The urea and uric acid are mixed with the water. This mixture is called **urine**. The urine is sent to your **bladder**. Your bladder is like a small sack. It stores the urine. When your bladder gets full, it empties the urine from the body through a special opening.

Some liquid wastes leave your body another way, too. When you are hot or exercise hard, you perspire. Liquid leaves your skin through small openings called **pores**. Perspiring helps cool you down.

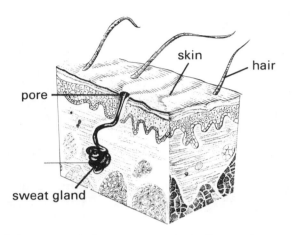

Try this #35

Materials: mirror
1. Take a deep breath.
2. Exhale onto the mirror.
3. What happened to the mirror? _____
4. Can you name this third way the body eliminates liquid waste?

_____

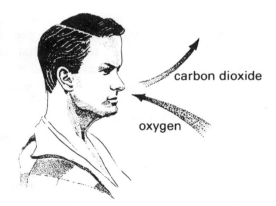

Elimination of gaseous wastes

When you breathe, you take in oxygen. Your cells use this oxygen. They give off carbon dioxide as a waste product. Your blood carries this carbon dioxide to your lungs. When you exhale, this carbon dioxide leaves your body.

Chapter checkup

Choose the best answer. Put your answer in the blank next to the question.

_____ 1. When animals breathe, they give off the waste product
   a. oxygen
   b. uric acid
   c. carbon dioxide

_____ 2. Another name for the colon is the
   a. small intestine
   b. large intestine
   c. anus

_____ 3. Urea is carried from the cells to the kidneys through the
   a. bladder
   b. bloodstream
   c. urine

_____ 4. In animals, the removal of wastes is called
   a. excretion
   b. perception
   c. digestion

_____ 5. Urine is stored in the
   a. large intestine
   b. pores
   c. bladder

_____ 6. Breathing removes from your system carbon dioxide and
   a. water
   b. solid wastes
   c. oil

_____ 7. Urea and uric acid are removed from the bloodstream by the
   a. red blood cells
   b. bladder
   c. kidneys

_____ 8. The ameba expels wastes through its
   a. cell membrane
   b. anus
   c. lungs

gland
hormones
pituitary
thyroid
adrenal
pancreas
insulin
gonads

23. The endocrine system

Imagine yourself walking through a forest. Suddenly you meet a very large, very grumpy bear. Your heart begins to pound very loudly and very quickly. You turn and run faster and farther than you've ever run before. You feel as if you're filled with unlimited enery.

Something has happened inside your body. You don't realize what has happened. Your endocrine system automatically mixed special substances in your blood. These substances helped you run faster and farther.

The endocrine system

The endocrine system is made up of many **glands**. A gland is an organ that makes and releases substances called **hormones**. Hormones are substances that help control activities of the body. The hormones are released into the blood. The blood carries them to where they are needed.

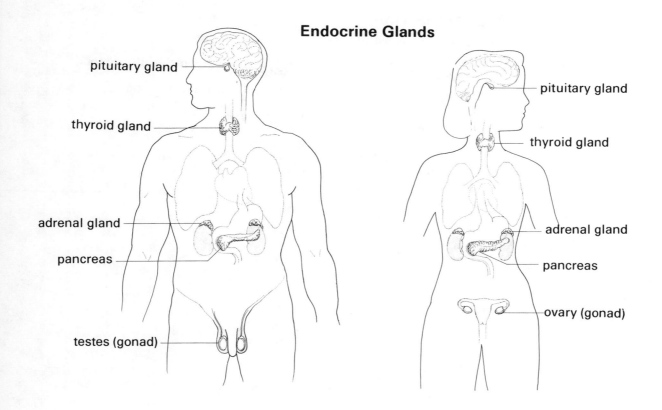

Endocrine Glands

pituitary gland

thyroid gland

adrenal gland

pancreas

testes (gonad)

pituitary gland

thyroid gland

adrenal gland

pancreas

ovary (gonad)

Different types of glands

The **pituitary gland** is a very important gland. It is located at the base of your brain. It is called the master gland. It produces special hormones. These hormones control many other glands. The pituitary gland also controls growth and reproduction.

The **thyroid gland** is in your neck. It determines how fast the cells use food. Sometimes, the thyroid gland works too much. You may eat a lot and still lose weight. Sometimes the thyroid produces too little. You may eat very little but gain a lot of weight.

The **adrenal glands** produce several different hormones. One hormone helps control the amount of salt and water in your body. Salt keeps your body fluids in balance. The adrenal gland releases other hormones when you become frightened or hurt. (It was your adrenal gland that kicked into action when you saw the bear.) The adrenals make a hormone that speeds up breathing, raises your blood pressure, and makes your body react to a scary or exciting situation. There are two adrenal glands—one on top of each kidney.

The **pancreas** is a gland that regulates the sugar level in the blood. It produces **insulin**. Blood carries sugar to the cells. Insulin allows your body to use the sugar. (The pancreas also is a digestive gland. It produces digestive enzymes.)

The **gonads** are called sex glands. They produce hormones that are important for reproduction. (See chapter 25.) The sex hormones stimulate the growth of body hair and muscles in males. Sex hormones stimulate the growth of breasts in females. The sex hormones help control sexual and reproductive activities.

Chapter checkup

Match the glands with the function they perform.

_____ 1. adrenal   a. controls the rate at which cells use food

_____ 2. gonads   b. helps a body react in an emergency

_____ 3. thyroid   c. the sex glands

_____ 4. pituitary   d. regulates sugar level

_____ 5. pancreas   e. controls growth

24. The senses

Humans and most other higher animals normally have five senses. They are touch, taste, smell, hearing, and sight. These senses allow the person or animal to receive information from the outside world.

Some senses may be stronger than others. Most dogs have poor eyesight, but they have a keen sense of smell. Some animals have special senses. Certain birds have a keen sense of direction. Bats have a special sonar that lets them find objects in the dark.

People who are blind or deaf usually develop their other senses to a high degree. A blind person may have very keen senses of hearing, smell, and touch to make up for the loss of eyesight.

Touch

The sense of touch tells you when you have come in contact with something. You feel heat, cold, pain, or pressure. Your skin is your sense organ. Nerve cell endings in your skin receive information. Different nerves register different feelings. There are cells that detect only heat. There are cells that detect only cold. Still other cells sense pressure. And other cells detect pain. The cells send the message to your brain (or spinal column in a reflex action). Your brain sends a message to the rest of your body on how to react.

Taste

Your tongue is your taste organ. It is covered with **taste buds**. Most tastes are a mixture of sweet, salty, sour, or bitter. The taste buds send messages to your brain about the tastes. Your brain often needs the help of the other senses to know what you are eating. You depend a lot on your senses of smell, sight, and touch when you eat.

Smell

Your nose is the organ for smelling. Nerve endings in your nose pick up odors. Other nerves carry the message to your brain.

Tongue

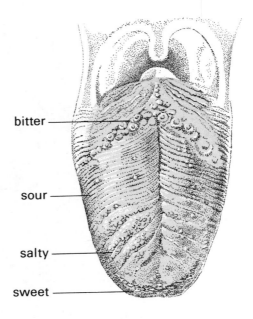

bitter ———

sour ———

salty ———

sweet ———

Try this #36

Materials: sugar, salt, lemon, saccharin dissolved in water
1. Press your finger into the sugar so that a few crystals cling to it.
2. Place the crystals on the tip of your tongue. Were you able to tell the sugar was sweet?
3. Place sugar crystals on the side of your tongue. Could you tell the sugar was sweet?
4. Place sugar on the back of your tongue. Could you tell the sugar was sweet?
5. Do the same thing for each of the substances listed above. Keep track of your results on a chart like the one below.

	Front	Side	Back
Sugar	————	————	————
Salt	————	————	————
Lemon	————	————	————
Saccharin	————	————	————

Were you able to tell which part of your tongue could identify each taste? Taste buds at the front of your tongue detect sweet or salty tastes. Taste buds at the sides of your tongue detect bitterness. Taste buds at the back detect sourness.

Materials: blindfold, knife, apple, carrot, pear, potato, and onion
1. Choose a partner.
2. Have one partner put on a blindfold.
3. Have the other partner cut off small pieces of the apple, carrot, pear, potato, and onion.
4. Have the blindfolded partner hold his nose and taste each of the foods. Have him try to identify each food as he tries it. Keep track in a *chart like the one below.*
5. If you'd like, switch roles so the other partner can try to identify the foods he or she is tasting.

apple	correct	incorrect	unsure
carrot			
pear			
potato			
onion			

Without smelling, most people aren't able to identify the different foods. When you eat, you use your senses of taste and smell to enjoy your food.

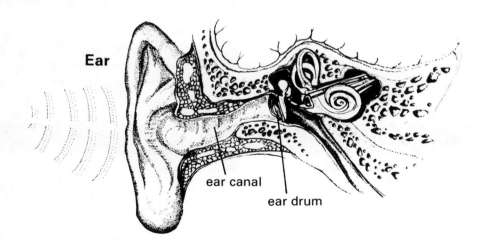

Ear

ear canal

ear drum

Hearing

Your ears are your sense organs for hearing. Your ears pick up sound waves. Sound waves are simply vibrating air. The air vibrations enter your ear and travel down your ear canal to your eardrum. The air vibrates against the drum. This causes tiny bones behind the eardrum to vibrate. This vibration sends the message on to the innermost part of your ear. The sound travels to a part of your ear that is filled with fluid and has nerve endings. The nerve endings receive the vibrations from the fluid. The nerve endings send the message to your brain to be interpreted.

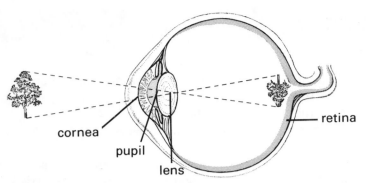

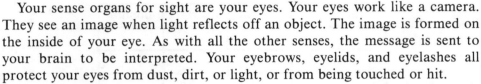

Light rays bend as they pass through the lens. The retina receives a reversed image. Nerve endings in the retina send the image to the brain to be interpreted.

Sight

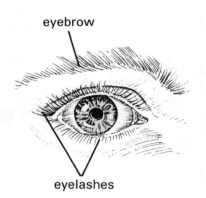

Your sense organs for sight are your eyes. Your eyes work like a camera. They see an image when light reflects off an object. The image is formed on the inside of your eye. As with all the other senses, the message is sent to your brain to be interpreted. Your eyebrows, eyelids, and eyelashes all protect your eyes from dust, dirt, or light, or from being touched or hit.

Light rays enter the clear **cornea** on the front of your eye. The cornea is part of the protective coating of your eye. The light passes into your **pupil**. The pupil is the dark opening in the middle of the colored part of your eye. Your pupil opens up wider if there isn't much light, in order to let more light in. It closes up smaller if there is a lot of light, so that too much light doesn't enter. The light passes from the pupil to the lens inside your eye. The light rays bend as they pass through the lens. Then they meet at a point on the **retina**. The retina is a thin layer of tissue on the inner eye. Nerve endings in the retina send the image to your brain.

A person is nearsighted or farsighted if images do not focus right on the retina. A nearsighted person sees clearly closeup, but he can't see distant objects clearly. A farsighted person sees distant objects clearly, but he can't see clearly closeup. Glasses or contact lenses correct these problems.

Try this #38

Materials: paper, pencil
1. Make two large periods about 5 inches apart on the paper.
2. Hold the paper about 8 inches from your face.
3. Close your right eye and stare at the period on the right side.
4. Move the paper slowly away from your face.
5. At what point did the period on the left side seem to disappear?

_____

6. Now close your left eye and stare at the period on the left side.

7. Did the period on the right seem to disappear? _____

There is a **blind spot** on your retina. There are no light-sensitive nerve cells on this spot on your retina. When the image inside your eye lands on your blind spot, you can't see the image.

Chapter checkup

Write **T** for true or **F** for false next to each statement below.

———— 1. In humans, the cells that detect hot do not detect pressure.

———— 2. Most higher animals have five senses.

———— 3. Most taste buds are located on the roof of your mouth.

———— 4. All five sense organs send messages to the brain.

———— 5. The taste buds that taste salty food are located on the back of your tongue.

———— 6. Your eyes see an image when light reflects off of it.

———— 7. Your ears hear by picking up sound waves.

———— 8. The pupil of your eye gets smaller if there isn't a lot of light.

———— 9. Nerve endings in the cornea send images to the brain.

fission	testes	vagina	fallopian tubes
budding	scrotum	uterus	menstruation
sperm	semen	cervix	embryo
fertilization	penis	ovaries	fetus

25. Reproduction

Methods of reproduction vary greatly among the different species of animals. Most lower forms of animals reproduce asexually. Asexual reproduction involves only one parent. Most higher forms of animals reproduce sexually. Sexual reproduction involves two parents.

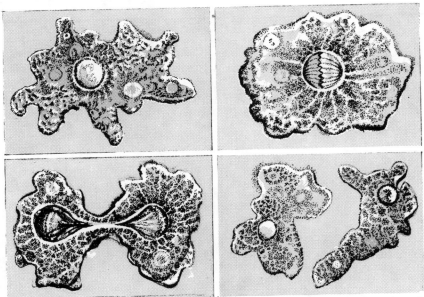

This ameba is reproducing by dividing in two. This is a form of asexual reproduction.

Asexual reproduction

The simplest method of asexual reproduction is called **fission**. Reproduction by fission is the splitting of one cell into two. A single-celled animal will divide itself into two cells. Each cell eventually becomes a separate organism.

Budding is another method of asexual reproduction. During budding, part of the parent's body separates. The part that broke away grows into a complete new animal. Sometimes the new organism remains attached to its parent. Or, it might separate and live on its own. Sponges are able to reproduce through budding.

In chapter 15 you learned that certain plants reproduce by spores. Some animals reproduce by spores also. Each spore is a single reproduction cell. Each spore can grow into a new organism.

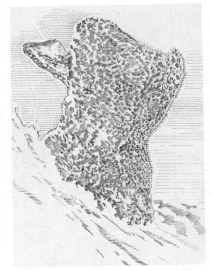

The bud on this sponge will break off from the parent sponge and grow into a new sponge.

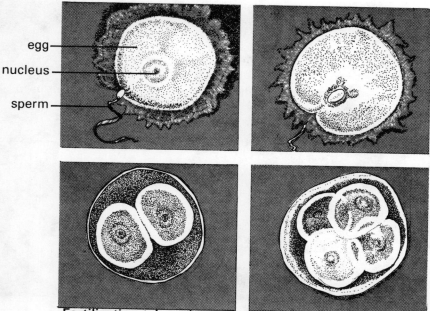

egg
nucleus
sperm

Fertilization takes place when sperm and egg cells unite. The fertilized cell divides many times and develops into an embryo.

Sexual reproduction

Sexual reproduction requires the mating of a male and female of a species. Mating is the uniting of the sex cells of the two parents.

Male sex cells are called **sperm**. A female sex cell is called an egg or ovum. During mating, one sperm unites with one egg. This is called **fertilization**. The fertilized egg divides many times. Through these cell divisions, a new animal develops.

Human reproduction

Humans (and other mammals) reproduce sexually through internal fertilization and development. The sperm fertilizes the egg while it is still inside the female. The fertilized egg develops inside the mother's body.

The male reproductive system. The sperm are produced in the **testes**. (The singular is testis. There are usually two.) The testes are inside the **scrotum**. After being made in the testes, the sperm are stored inside the male's body. During sex, the male expels **semen**. Semen is a liquid that has the sperm in it. The semen comes out through the **penis**, the main sex organ of the male.

The female reproductive system. The **vagina** is the passageway to the **uterus**. The uterus is a pear-shaped organ. If an egg is fertilized, it will develop inside the uterus for nine months. The **cervix** is at the opening to the uterus. Eggs are produced inside the **ovaries**. Each month (for most women), an egg leaves an ovary. It travels through the **fallopian tube** to the uterus.

Each month, the uterus prepares for a fertilized egg. The lining of the uterus becomes rich with blood and nutrients. If no egg is fertilized, the lining is shed, along with the unfertilized egg. That is **menstruation**. For most women, this happens monthly. It is called the menstrual period.

Conception. Conception is the moment when the sperm fertilizes the egg. During intercourse, the male puts his penis inside the woman's vagina. He releases semen into the vagina. The sperm are in the liquid semen. The sperm swim to the egg. About 350 million sperm are released at a time. Only a few make it to the fallopian tube. It is here that fertilization may occur if there is an egg on its way down.

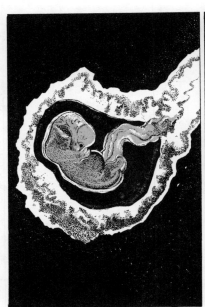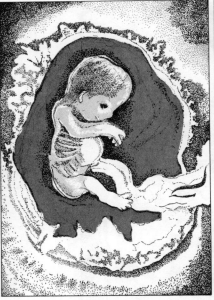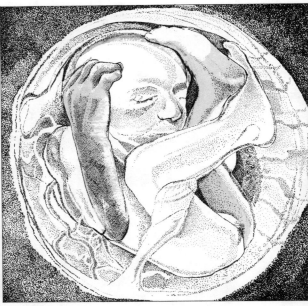

Development of a Human Fetus

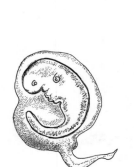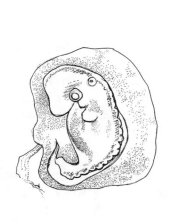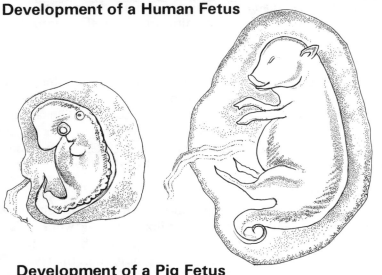

Development of a Pig Fetus

Intercourse is the way most pregnancies begin. But some women become pregnant through artificial insemination. During artificial insemination, semen is injected into the woman's vagina through a tube or large needle. There have even been "test tube babies." The egg is fertilized outside the mother's body in a laboratory. The fertilized egg is then implanted in the mother's uterus.

Development. If an egg is fertilized, it starts to divide. In about a week, this ball of cells plants itself in the uterus' wall, which is thickened with blood and nutrients.

It will take about nine months from conception for a human to develop inside its mother's body. For the first three months, the developing being is called an **embryo**. During this time, many of the internal organs are formed. After three months, the developing being is called a **fetus**. After about nine months, the baby will be pushed out through the vagina of the mother. It is now ready to survive outside its mother's body.

116

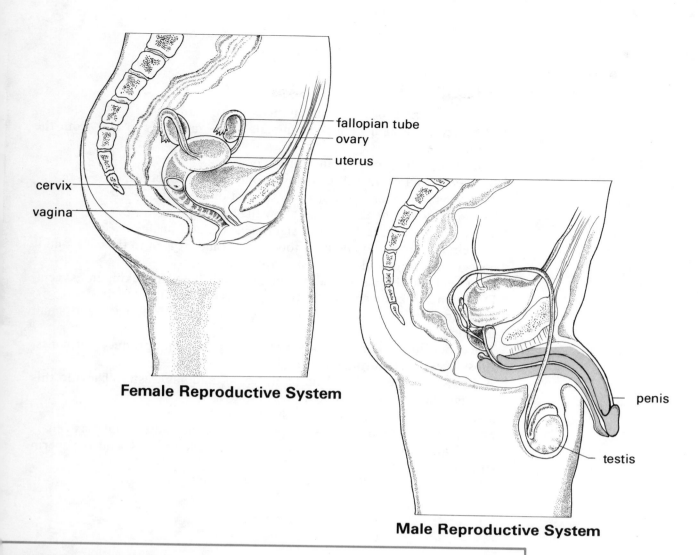

Female Reproductive System

fallopian tube
ovary
uterus
cervix
vagina

Male Reproductive System

penis
testis

Chapter checkup

Write **T** for true or **F** for false next to each statement.

_____ 1. During the first three months a developing being is called a fetus.

_____ 2. Most higher animals reproduce asexually.

_____ 3. Budding is a form of asexual reproduction.

_____ 4. Eggs are produced in the testes.

_____ 5. The union of a sperm and an egg is called fertilization.

_____ 6. The male sex cell is called a sperm.

_____ 7. During budding, part of the parent's body separates.

_____ 8. Spores develop as a result of the union of a sperm and an egg.

_____ 9. Asexual reproduction involves only one parent cell.

_____ 10. The embryo develops inside the uterus.

_____ 11. Menstruation is the shedding of the enriched lining of the uterus.

Unit review

- The life functions of plants and animals are similar.
- The nervous system of higher animals sends messages to and from the brain.
- The brain is the control center for the body.
- Messages from nerve cells can be sent only in one direction.
- The skeleton supports and protects the rest of the body.
- The muscles are the tissues that allow the body to move.
- The five senses are taste, touch, sight, hearing, and smell.
- The circulatory system delivers food, water, and oxygen to the cells, and it collects carbon dioxide and wastes from the cells.
- Most land vertebrates, including humans, have lungs for taking in oxygen.
- Digestion breaks down food so that the cells can use it.
- Digestion begins in the mouth and continues in the stomach and small intestine.
- The endocrine system is made up of many glands that produce hormones, substances that control body functions.
- Animals must get rid of their solid, liquid, and gas wastes. They do this through their excretory system .
- Asexual reproduction requires only one parent.
- Sexual reproduction requires the mating of a male and a female sex cell.
- Fertilization is the joining of the egg (the female sex cell) and the sperm (the male sex cell).
- Humans and other mammals develop inside the mother's uterus.

Unit checkup

Fill in the blank

1. People who study animals are called _____.

2. _____ is the process of changing food to a usable form.

3. The system for transporting food and water is called the _____ system.

4. The elimination of all wastes is called _____.

5. Hormones are made in the _____ system.

6. The brain is the control center of the _____ system.

7. The _____ stores urine until it leaves the body.

8. The two _____ filter your blood.

9. Undigested parts of food leave the small intestine and enter the _____.

Multiple choice

1. A long tube that leads to the stomach
   a. colon
   b. esophagus
   c. ligament

2. Chemicals that help digest food
   a. enzymes
   b. hormones
   c. plasma

3. A liquid in your mouth that helps start the digestive process
   a. saliva
   b. hormones
   c. taste buds

4. A special acid produced by your stomach for digestion
   a. antacid
   b. hydrochloric acid
   c. uric acid

5. Short branches coming out of a nerve cell
   a. capillaries
   b. dendrites
   c. axons

6. The part of the brain that controls thinking, learning, and talking
   a. cerebellum
   b. cerebum
   c. thyroid

7. The part of the brain that controls the muscles
   a. medulla
   b. spinal cord
   c. cerebellum

8. The part of the brain that controls the heart and lungs
   a. cerebellum
   b. cerebrum
   c. medulla

9. The route that all information takes to and from the brain
   a. pancreas
   b. spinal cord
   c. artery

True or false

Write **T** or **F** next to each statement. If the statement is false, correct it so that it is true. Write a *complete* sentence.

_____ 1. Plasma is the solid part of blood.

_____

_____ 2. Red blood cells fight disease.

_____

_____ 3. Platelets cause blood to clot.

_____

_____ 4. Arteries carry blood to the heart.

_____

_____ 5. The aorta is the largest artery in the body.

_____

Unit 5.

It's in your genes

Every plant and animal receives traits from its parent or parents. Inside an apple there are seeds. Inside the seeds, there is a coded message that tells the seeds to grow into another apple tree.

Your parents passed on traits to you. Inside the sperm and egg, there were coded messages. The messages said you would be a certain color and height. The message said what color your hair and eyes would be. The message said you would have a certain blood type and that your nose, and mouth, and hands would be a certain shape. In this unit, you will find out how you came to look like you—and why you might look so much like one of your relatives.

Gail Denham

Robert Winslow

Jan Doyle

USDA

26. Heredity

Dwight R. Kuhn

Pea plants were used to learn how traits are passed from parent to offspring.

"She has her father's eyes." "He has his mother's nose." Have you ever heard people say things like that? Children inherit many of the traits of their parents. **Heredity** means the passing of traits from parents to offspring. **Genetics** is the study of heredity.

Mendel's experiments

The first person known to study heredity was an Austrian monk named Gregor Mendel. In 1866, he published his experiments with pea plants. Pea plants reproduce sexually. Pollen from one plant fertilizes an ovule (egg) from another plant. The offspring have traits of both parent plants.

Mendel worked with yellow and green peas. First he took the pollen from a yellow plant. With it he fertilized the ovule of another yellow plant. As expected, the offspring were yellow. We can show that with the chart below. The yellow trait is represented by a capital Y.

The offspring have traits of both parents. In this case, the traits are the same, YY.

	Y	**Y**
Y	YY	YY
Y	YY	YY

Mendel did the same thing with green plants. He took the pollen from a green plant. With it, he fertilized the ovule of another green plant. As expected, the offspring were green. We will use a lower case y to stand for the green trait.

	y	y
y	yy	yy
y	yy	yy

Mendel then crossed a yellow pea plant with a green pea plant. He expected half the offspring to be green and half to be yellow. He made a chart like the one below.

	Y	Y
y	Yy	Yy
y	Yy	Yy

The offspring have both green and yellow traits. But all the plants were yellow. Mendel decided that the yellow trait must be stronger than the green trait. He called the stronger trait dominant. In this case, yellow was the **dominant trait**. He called the weaker trait recessive. In this case, green was the **recessive trait**.

Mendel then took the offspring from the experiment he had just done and mated them. This time, a curious thing happened. One out of every four plants was green. He again set up a chart. Remember from the previous chart that he bred plants with Yy traits.

	Y	y
Y	YY	Yy
y	Yy	yy

Some of the offspring were pure Y (YY). They, of course, were yellow. Some of the plants had both Y and y traits. Because yellow is the dominant trait, these offspring were yellow also. But some of the plants were pure green (yy). These plants did not carry the yellow trait. The chart shows that one out of every four plants should be green.

Things to do

Complete the charts and answer the questions.
B stands for brown hair, a dominant trait.
b stands for red hair, a recessive trait.

1. A man with Bb traits and a woman with Bb traits have four children.

	B	**b**
B	BB	Bb
b	Bb	bb

a. How many of their children should have brown hair? _____

b. How many of their children should have red hair? _____

2. A man with BB traits and a woman with Bb traits have four children.

	B	**B**
B	BB	BB
b	Bb	Bb

a. How many of their children should have brown hair? _____

b. How many of their children should have red hair? _____

3. A man with bb traits and a woman with bb traits have four children.

	b	**b**
b	bb	bb
b	bb	bb

a. How many of their children should have brown hair? _____

b. How many of their children should have red hair? _____

124

Materials: none
1. Look at the picture below. Try to roll your tongue as the girl is doing.

2. How many people in your class can roll their tongues? _____

3. How many people in your class cannot roll their tongues? _____

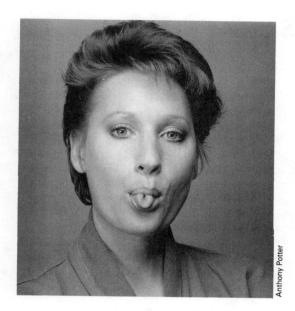

Anthony Potter

In humans, the ability to roll your tongue is a dominant trait. On the average, for every three people who can roll their tongue, one person cannot. Does your class come close to the average?

Chapter checkup

Circle the correct answer.

1. The *dominant, recessive* trait is the strong trait.
2. *Offspring, Genetics,* is the study of heredity.
3. The *dominant, recessive* trait will appear in the offspring only if both parents carry that trait.
4. The first person to study heredity was *Mendel, Einstein.*
5. The passing of traits to offspring is called *heredity, donation.*
6. Mendel first experimented with *pea, tomato* plants.

27. Chromosomes

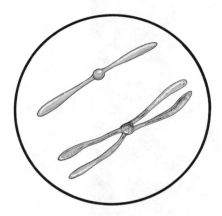

Chromosomes

Chromosomes are the magic in heredity. They are the material in the cells that carries the organism's traits. The chromosomes of Mendel's pea plants determined what traits the plants would have. If you have blue eyes, it is because your chromosomes carry the trait for blue eyes. Your body has more than 40 trillion cells. Every cell contains chromosomes. All living organisms have chromosomes that determine what traits will be present.

Different organisms have a different number of chromosomes. Chromosomes come in pairs. The cells in a fruit fly all have four pairs of chromosomes. That means each cell has eight chromosomes. The cells in a piece of corn each have 10 pairs of chromosomes. That means each cell has 20 chromosomes. The cells in your body have 23 pairs of chromosomes. That means each cell has 46 chromosomes.

Mitosis

Your body is constantly making new cells. It does this by mitosis. **Mitosis** is cell division. As the cell begins to divide, the chromosomes split lengthwise and make copies of themselves. They go to opposite ends of the cell. Then the cell divides. Each new cell has a copy of every chromosome. Each nucleus again has 23 pairs of chromosomes, or 46 chromosomes. The two cells look alike.

The chromosomes carry all the information about that cell. A blood cell will always divide into other blood cells, for example. As the cell divides, each new cell receives a complete set of chromosomes.

Here are the six stages of mitosis.

1. As mitosis begins, each chromosome splits and each half makes an exact copy of itself.

2. The membrane that surrounds the nucleus disappears. The nucleus blends with the cytoplasm.

3. The pairs of chromosomes line up in the center of the cell.

4. The pairs of chromosomes grow apart. One member of each pair will go to one side of the cell. The second chromosome of each pair will go to the other side of the cell.

5. A new nucleus will form around each group of chromosomes.

6. The cytoplasm splits, forming two complete cells. Each cell is exactly like the cell it came from.

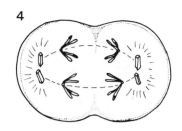

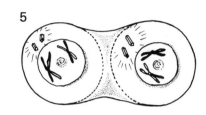
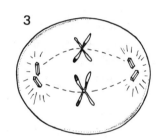

Try this #40

Materials: microscope, pre-made onion slide
1. Observe the onion slide under low power.
2. Do you see any cells in the stages of dividing?
3. Draw what you see.
4. Do the same under high power.
5. Draw what you see.

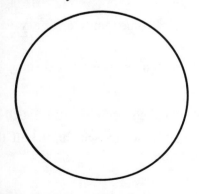
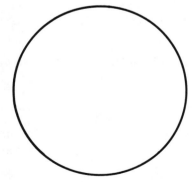

What you saw were cells in the various stages of mitosis. Each cell divides and grows many, many times. This is what made the onion grow. Did you recognize any of the stages of mitosis?

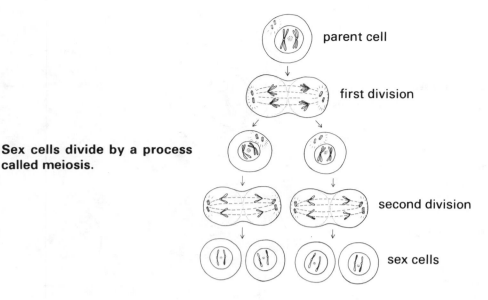

parent cell

first division

Sex cells divide by a process called meiosis.

second division

sex cells

Meiosis

Meiosis is the division of sex cells. Sperm are the sex cells of males. Eggs are the sex cells of females. Like all other cells, sex cells divide. But sex cell division is slightly different from the division of other kinds of cells. Sex cells go through two divisions. After the first division, there are two cells. Each still has 23 *pairs* of chromosomes. But after the second division, each sex cell has 23 chromosomes rather than 23 *pairs* of chromosomes. The sperm cell has 23 chromosomes. The egg cell has 23 chromosomes. Now, if the egg is fertilized by the sperm, the fertilized egg will once again have 46 chromosomes: 23 chromosomes from the sperm (the father) and 23 chromosomes from the egg (the mother). The offspring will have traits of both parents because it has chromosomes from both parents. The fertilized egg will divide over and over again through mitosis. Each cell will have 23 pairs of chromosomes.

Chapter checkup

Write **T** for true or **F** for false in the blanks below.

_____ 1. A human cell has 46 pairs of chromosomes.

_____ 2. Sex cells have half the number of chromosomes ordinary cells have.

_____ 3. All the traits of a child are determined by the chromosomes it receives from its mother.

_____ 4. Cells, other than sex cells, divide by a process called mitosis.

_____ 5. When a cell divides by mitosis, each new cell will have half the number of chromosomes as the original cell.

_____ 6. Sex cells divide by a process called meiosis.

_____ 7. All the traits of an organism are carried in its chromosomes.

_____ 8. All organisms have exactly 46 chromosomes.

28. Genes and DNA

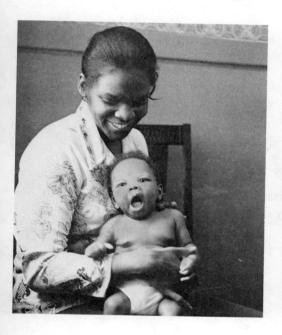
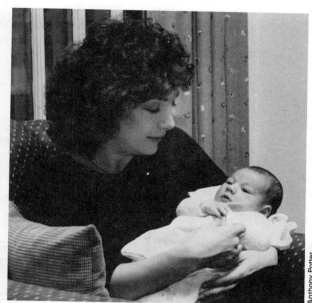

Look at yourself. It is amazing how your body is made up of many different parts. Look at the shape of your eyes and ears. Do you have freckles? How big are your feet? Every trait is determined by the chromosomes given to you by your parents. You have only 46 chromosomes. How can all these traits be determined by only 46 chromosomes? Let's look more closely at chromosomes and find out.

Genes

All cells have chromosomes. You studied chromosomes in chapter 27. Chromosomes look like long pieces of thread. On each thread are hundreds of **genes**. Each gene is responsible for a different part of you. There are genes that determine your hair color. There are genes that determine the shape of your nose. Genes are responsible for all your traits. The study of genetics is really the study of genes. The word *genetics* comes from the word *gene*.

You started out as a single cell, a fertilized egg. Half the genes in that cell came from your father. Half came from your mother. Through mitosis, that one cell split into two cells. The genes in the two new cells were exactly like the genes in the first cell. The two cells divided into four cells. Again, the genes in each of the four cells were exactly like the genes in the original cell.

How genes duplicate themselves

Each gene is made of a complex molecule called **DNA**. The DNA molecule looks like a twisted rope ladder. It is the basic genetic material. When a cell splits, the chromosomes are splitting. Inside the chromosomes, there is also a lot going on. The DNA is splitting.

During mitosis, the DNA molecule begins to unwind. As it unwinds, it splits down the middle.

Each strand begins to rebuild the missing side from the chemicals found in the cell. Each new part must be exactly like the old one.

When it splits entirely, two new DNA molecules are formed. Each is exactly like the original.

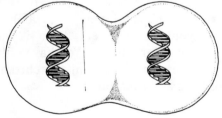

As mitosis goes on, one DNA molecule goes to one side of the cell. The other DNA molecule goes to the other side of the cell.

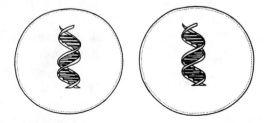

When the cell splits, each new cell has a DNA molecule exactly like the original.

Chapter checkup

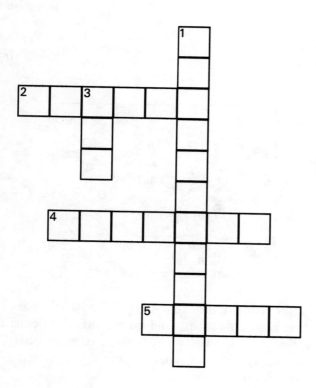

Down

1. Threadlike structures that determine all your traits
3. The name of the molecule that makes up the genes

Across

2. The DNA molecule looks like a _____.
4. The process by which molecules divide
5. Molecules on the chromosomes; each one is responsible for a different trait.

Science words

selective breeding
strains
crossbreeding
inbreeding
clone

29. Using genetics

" I HATE THESE FAMILY REUNIONS"

Farmer Brown raises beans. Every year she sells most of what she grows. But she saves the best beans for seeds. She knows the best beans will grow into the best plants. She is using the laws of genetics to grow better beans.

Scientists have used many different types of genetic "engineering" to make changes in plants and animals.

Breeders chose the parents of this championship horse. The horse inherited traits from its parents.

USDA

Selective breeding

In the example above, Farmer Brown used **selective breeding**. In selective breeding, only those plants with the most desirable traits are used to produce the next generation. Farmer Brown used only the best seeds.

This method is also used in breeding animals. Breeders choose parents with the most desirable traits. The breeders want the offspring to inherit the most desirable traits. Championship race horses are often bred this way.

Crossbreeding

Strains are different varieties of the same species. In chapter 26, you read about Gregor Mendel's pea plant experiments. He used different strains of peas. One strain was yellow. The other strain was green. In **crossbreeding**, two different strains are bred together. Mendel crossed the green strain with the yellow strain. He used crossbreeding to produce a new strain with both green and yellow traits.

Crossbreeding is often used for breeding cattle. One strain of cattle was very resistant to heat, insects, and disease. Another strain of cattle was an excellent beef animal. When these two types of cattle were bred together, the offspring were good beef animals that were also able to resist heat, insects, and disease.

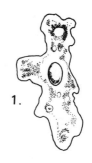

1.

Inbreeding

Inbreeding is the mating of animals of the same or closely related stock. Since both parents are genetically similar, their similar traits will be passed to the offspring.

Clones

A **clone** is an organism that is produced asexually from a single cell. It is genetically identical to the parent cell. Clones are the result of asexual reproduction. Asexual reproduction involves only one parent; there is no sperm or egg. The parent cell divides. Any offspring will have the same genetic makeup as its parents. Thus the parents and the offspring are clones. Most single-celled organisms reproduce by mitosis (cell division) and are clones.

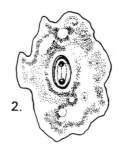

2.

Clones in the laboratory

Frogs only reproduce sexually. Frogs cannot naturally produce clones. But in the laboratory, scientists have been able to artificially produce frog clones. The method is very complicated. It involves removing the nucleus of a frog egg. It is replaced by the nucleus of a body cell from another frog. The new nucleus has a complete set of genes. A new frog grows from this single cell. The resulting offspring is a clone of the second frog since the genes are identical.

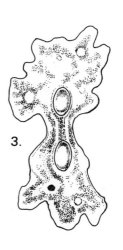

3.

Chapter checkup

Circle the right answer.

1. A *strain, clone* is an organism that is genetically identical to another organism.

2. Clones are the result of *asexual, sexual* reproduction.

3. Farmer Brown used *selective breeding, crossbreeding* when she saved only the best seeds for next year's planting.

4. Self-pollination is a form of *asexual reproduction, inbreeding.*

5. In *inbreeding, selective breeding,* two different strains are bred together.

6. *Sperm, Strains* are different varieties of the same species.

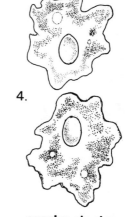

4.

ameba cloning

133

Unit review

- Genetics is the study of heredity, the passing on of traits.
- Traits are passed from parents to offspring.
- The sperm carries the traits of the father.
- The egg carries the traits of the mother.
- Chromosomes are the cell material that carry the organism's traits.
- Chromosomes are made up of hundreds of molecules called genes.
- Each gene is responsible for a different trait.
- Genes are made up of a complex molecule called DNA. DNA is the basic genetic material.
- Cells (except sex cells) divide by a process called mitosis.
- Sex cells divide by a process called meiosis.
- For organisms that reproduce sexually, offspring receive half their chromosomes from the mother and half from the father.

Unit checkup

A. Understanding what you have read

Write *agree* or *disagree* for each statement below. If you disagree, write a complete sentence that corrects the statement.

1. Children do not inherit many traits from their parents.

   _____

2. Gregor Mendel was a German monk who studied human reproduction.

   _____

3. Humans have 47 chromosomes.

   _____

4. DNA molecules look like small dots.

   _____

5. Human sperm and egg cells have 46 chromosomes each.

   _____

B. Fill in the blank

1. Genetics if the study of _____.
   a) heredity   b) meiosis   c) environment

2. A dominant trait is _____ than a recessive trait.
   a) weaker   b) stronger   c) more colorful

3. _____ are responsible for all your traits.
   a) foods   b) genes   c) red blood cells

4. DNA looks like a _____.
   a) twisted rope ladder   b) web   c) circles

5. Division of sex cells is called _____.
   a) meiosis   b) mitosis   c) reproduction

Complete the charts

R stands for red blossoms; red is a dominant trait.

r stands for white blossoms; white is a recessive trait.

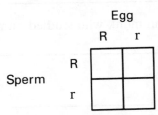

1. How many of the four offspring are red? _____

2. How many are white? _____

3. How many of the four offspring are red? _____

4. How many are white? _____

Mitosis

Number the steps of mitosis to put them in the proper order.

_____ a. A new nucleus will form around each group of chromosomes.

_____ b. The membrane around the nucleus disappears.

_____ c. Each chromosome splits and makes an exact copy of itself.

_____ d. The cytoplasm splits, forming two complete cells.

_____ e. The pairs of chromosomes line up in the center of the cell.

_____ f. Chromosomes and their copies line up on opposite sides of the cell.

Unit 6.
Health is good for you

Many people say good health means not being sick. But good health is more than that. Good health means our whole self is at its best—physically and emotionally and mentally. To be healthy, we have to eat right, exercise, know how to relax, and be informed about substances or activities that can be harmful. You have been learning about all the systems of your body. Your body is a miracle. Take good care of it. Take good care of you.

contagious
epidemic
antibodies
immune system
vaccine
addiction

30. Diseases

In 1908, Captain Donald MacMillan made one of many trips to the cold arctic. On one of his voyages, he met a tribe of Eskimos. These Eskimos had never met anyone outside of their tribe before.

None of these Eskimos had ever had a cold. After they met these outsiders, many of the Eskimos caught colds. Colds then spread throughout the entire tribe. Ever since then, these Eskimos have been catching colds.

Colds are caused by micro-organisms called viruses.

Things to do

1. Why do you think the Eskimos never caught colds before meeting the outsiders? _____

2. What do you think caused them to catch colds after meeting the outsiders? _____

What is a disease?

Many diseases are caused by micro-organisms. We often call them germs. Doctors call them bacteria and viruses. Germs are in food and water. They are on almost everything you touch. They are even in the air.

Different germs cause different diseases. Every year many people get the flu. The flu is caused by a flu virus. There are many different kinds of flu viruses. Most diseases caused by germs are **contagious**. Contagious means they can be passed from one person to another. Sometimes contagious diseases spread very quickly. Many people in an area catch them. This is called an **epidemic**.

138

Some common germ-caused diseases

Disease	Symptom	Preventative measures
Chicken pox	Headache, fever, skin rashes	None—having the disease prevents having it again
Flu	Fever, chills, aches and pains	Flu shot
Diphtheria	Sore throat, fever, hoarseness	Diphtheria shot
Measles	Fever, aches, cough, skin rashes	Measles shot

Not all diseases are caused by germs. These other diseases are not contagious. Some are caused by disorders of the cells or tissues. Cancer is a disorder of the cells. Cancer causes cells to begin to multiply out of control. Some diseases are caused by birth defects. A poor diet can cause a disease called scurvy. Air and water pollution can cause diseases. Many coal miners develop a disease called black lung from breathing in coal dust. Even stress or tension can cause disease. Headaches and ulcers are often the result of too much stress.

Your body protects you

Your body is your best protection against disease. Your skin and eyes stop most foreign germs from entering your body. Your nose and throat have special linings that trap and kill germs. Coughing and sneezing eject foreign substances and germs from your mouth, nose and throat. Special cells in your lungs kill and eat germs. Your normal body temperature kills many types of germs. Your body raises or lowers its temperature to kill germs that are not killed by your normal temperature. Your white blood cells kill germs in your bloodstream.

Your body also makes special chemicals called **antibodies**. Antibodies fight foreign substances that have entered your body.

Antibodies throughout the body fight foreign substances and germs

White blood cells throughout the body kill germs

Body temperature kills many types of germs.

Lymph flows through the body and helps fight infection.

Most people have an effective **immune system**. The immune system helps them fight off disease. Some immunity is natural. That is, people are born with the ability to fight off some germs. Other immunity is built up. Babies get many illnesses. Their immune system is developing. Sometimes getting sick is a way to build up immunity. Getting sick prompts the body to make antibodies to fight the disease. The antibodies build up in the system. The person is able to fight off the disease next time he or she is exposed to it. Very few people get chicken pox or mumps twice, for example.

Some immunity is built up by the use of vaccines. A vaccine is a substance that contains dead or weakened germs that cause a disease. Your blood then forms antibodies against that disease. Then, if you are exposed to the disease, you probably won't catch it. Or, you will get a very mild case of it. Some diseases almost have been wiped out because of vaccines.

A few people are born with faulty immune systems. Their bodies cannot fight disease. Even a cold can be a killer.

AIDS is a disease that causes the immune system to stop working right. AIDS stands for "Acquired Immune Deficiency Syndrome." If someone gets AIDS, his or her body can no longer fight infection and disease. AIDS is spread through the transmission of body fluids, usually sexual contact or by sharing needles to take drugs. Scientists are trying to learn more about AIDS and how to care for people with AIDS.

SURGEON GENERAL'S WARNING: Smoking Causes Lung Cancer, Heart Disease, Emphysema, And May Complicate Pregnancy.

Anthony Potter

Tobacco, drugs, and alcohol

Drugs, alcohol, and tobacco can all cause serious damage to your body. Alcohol and tobacco are actually kinds of drugs. Drugs are chemicals that affect your body.

You can become addicted to these substances. Addicted means dependent on. If you are addicted, you crave the substance. If you don't have it, you feel sick. But taking it can make you sick, too. An **addiction** is an illness. Drugs alter how you look at the world and how you interact with friends and relatives. Thinking about how to get the substance can eat up a lot of your time and energy.

Tobacco. Long-term smoking can cause cancer, or lung and heart disease. Even young smokers may have shortness of breath, coughing, and other breathing problems. Smokeless tobacco is snuff and chewing tobacco. Smokeless tobacco can cause the same problems as cigarettes. Smokeless tobacco also can cause mouth cancer and gum disease.

Nicotine is the substance in tobacco that people can become addicted to. Nicotine is a stimulant. It makes the heart beat faster. It raises the blood pressure. The tar in tobacco contains many cancer-causing substances.

Secondhand smoke is smoke that is exhaled by a smoker. There is more and more evidence that secondhand smoke causes health problems for nonsmokers. It is polluting the air.

Drugs. There are many kinds of drugs—marijuana, sleeping pills, hallucinogens, heroin, cocaine, crack, alcohol, and medicines. All these drugs have different effects. They affect you physically, mentally, and emotionally. You can become addicted to drugs. Many can cause lasting damage to your brain, heart, and nervous system. They can cause birth defects. Sharing needles for taking drugs can also spread diseases like AIDS. Some drugs or combinations of drugs can kill.

Did you know that caffeine is a drug? Caffeine is in coffee, tea, and many soft drinks. Too much caffeine can cause a fast heartbeat, sleeplessness, and jitters. People can become addicted to caffeine.

Alcohol. Alcohol can dull your senses and affect your judgment. It may make you clumsy. Drinking too much for many years can cause liver, heart, and brain damage. Drinking too much can shorten your life. Alcoholism is addiction to alcohol. Like addiction to other drugs, alcoholism affects your ability to work and to interact with other people. It touches all the people around you.

Drugs can cause physical and emotional diseases. But these diseases can be prevented by avoiding drugs.

Chapter checkup

Place the letter of the correct answer in each blank below.

a. germ

_____ 1. Can be passed from one person to another

b. epidemic

_____ 2. A general illness of the body

c. disease

_____ 3. Disease that makes cells multiply out of control

d. addiction

_____ 4. Common name for disease-causing micro-organism

e. contagious

_____ 5. Disease spread rapidly throughout an area

f. scurvy

_____ 6. A disease caused by a poor diet

g. cancer

_____ 7. Dependence on a drug

h. antibody

_____ 8. Chemical in your body that fights disease

i. drug

_____ 9. Chemical you take that affects your body

nutrients
nutrition
minerals
carbohydrates
fats
proteins
vitamins
processed foods

31. Nutrition

Anthony Potter

Foods contain **nutrients**. Nutrients are the substances that promote growth and energy. Nutrients are used in your body to replace dead cells. The nutrients aid in repairing damaged tissue. They supply your body with the energy it uses every day.

Nutrition is the process your body uses to absorb essential nutrients. Nutrition affects your physical and emotional health. The nutrients your body needs are divided into six groups: water, minerals, carbohydrates, fats, proteins, and vitamins.

Water has no nutritional value. Yet, it is called a nutrient because it is so important to your body. Your body is mainly water. If your weigh 150 pounds, about 100 pounds is water. Water is needed in your cells. It helps bring digested foods to your cells. It helps get rid of wastes from your body. Most of the foods you eat contain water. You should drink about five glasses of water every day.

Minerals are body-building nutrients. Minerals like calcium and phosphorus help build and strengthen teeth and bones. Calcium and magnesium help your nerves and muscles work. Iron is needed by your red blood cells. Your body also needs iodine, cobalt, manganese, zinc, sulfur, sodium, chlorine, potassium, and copper.

Carbohydrates are energy nutrients. They help supply the energy you need every day. More than half of your diet probably is carbohydrates.

Sugar and starch are two types of carbohydrates. There is carbohydrate sugar in fruits, honey, and molasses. White or refined sugar is found in a lot of prepared foods. It has little nutritional value. Vegetables are a source of carbohydrate starches. Grains are full of carbohydrates. (Grains include rice, wheat, corn, and oats.) Most raw carbohydrate foods are also high in fiber. Fiber is also called roughage. It helps your body get rid of waste. Raw fruits and vegetables are fiber foods.

Fats are energy nutrients, too. But fats give you more than twice the calories that carbohydrates give you. Fats are found in such foods as butter, cream, cheese, shortenings, oils, nuts, and meats.

You must watch the amount of carbohydrates and fats you eat. If you eat more than your body needs, the extra will be stored as body fat. you will become overweight. Being overweight is bad for your health.

Proteins help you grow. Proteins also help your body repair itself. Proteins are found in meats, eggs, milk, cheese, nuts, and dry beans.

Vitamins are necessary for growth and body activity. They also help prevent diseases. There are many different types of vitamins. Some vitamins are referred to by letters. Vitamin A is found in most meats, butter, eggs, green beans, and many other foods. Vitamin A helps your body grow. It also helps keep you strong.

There are several B vitamins. They are called the vitamin B complex. They include vitamin B_1, vitamin B_2 and so on. The B vitamins help prevent disease, help your growth, and help your heart. The B vitamins are found in meats, eggs, milk, and cereals.

Vitamin C helps prevent diseases. It also helps build strong teeth and gums. It is needed in the blood cells to keep them strong. Vitamin C is found in oranges, tomatoes, leafy green vegetables, and other raw vegetables.

Vitamin D is the only vitamin that your body can make. Your skin can make some vitamin D. But your skin cannot make all the vitamin D that you need. You must get the rest from foods like butter, cream, and eggs. Sometimes milk has vitamin D added to it. Vitamin D helps prevent disease. It is also needed for strong bones and teeth.

Substance	Essential for	Source
Water	Cells, digestion, getting rid of waste	All foods, water
Minerals	Blood, bones, heart, nerves, muscles	Milk, meats, whole-grain cereals, vegetables, seafood
Carbohydrates	Energy	Fruits, vegetables, grains
Fats	Energy	Butter, cream, cheese, oils, nuts, meats
Proteins	Growth, body repair	Lean meats, eggs, milk, wheat, beans, peas, cheese
Vitamins	Growth, body activity, disease prevention	Most foods, especially milk, butter, meats, fruits, vegetables

Try this #41

Materials: vinegar, 1 jar, chicken bone
1. Fill the jar with vinegar.
2. Place the chicken bone in the jar.
3. Wait five days. Remove the bone from the jar.
4. Look at the bone carefully. Try to bend it.
5. What happened to the bone that was in the vinegar? Has it changed since you put it into the jar?

___

The vinegar dissolved the calcium and phosphorus out of the bone. Without the calcium and phosphorus the bone became soft. Without these two substances in your diet, your bones would become soft. They would not be able to support your body. You can see how important minerals are to you.

Eating right

Nutritionists are experts on eating right. Many nutritionists disagree on what "eating right" means.

For a long time, nutritionists have divided food into four main groups. They are milk, meat, bread and cereal, and vegetables and fruits. Most nutritionists say that a balanced diet includes food from each group every day. They say not to eat too much of any one kind of food.

Some experts add this warning. They say we eat too many **processed foods**, instead of fresh foods. Processed foods are canned or frozen, and they often have chemicals and coloring added. Most do not look much like the raw food they were made from. These experts say that many of the nutrients have been lost in the processing. They say that the added chemicals and coloring are not good for us. Nutritionists suggest we eat more fresh foods.

There is another group of nutritionists that says "eating right" may just depend on who we are. That is, different ethnic groups may have different needs. People of other nations have very different diets than we do. People in China eat very little dairy food, for example.

Nutritionists and doctors also disagree on whether to take vitamin pills. Some say the four food groups give us all the vitamins and minerals we need. Others say we do not get the needed amount in our diets or that we may even need extra.

To eat wisely...
— Eat a variety of foods.
— Eat when you're hungry, not when it's time or because you are bored.
— Eat fresh foods.
— Don't eat a lot of junk food and food with a lot of refined sugar.
— Pay attention to how you feel after you eat something. Do you feel satisfied, or do you feel stuffed? Do you feel sick from that kind of food? (Maybe you should avoid it then.)

Things to do

Bring in some packaged foods from home. Read the labels. The ingredients are listed in order of amount. The first item is the main ingredient. Are you surprised by anything you found on the labels? Discuss your findings with your class.

Anthony Potter

Chapter checkup

A. Fill in the blanks with the following words:

water   fats
minerals   proteins
carbohydrates   vitamins

1. _____ are needed for growth and body activity and to help in preventing disease.

2. _____ are the energy nutrients that supply twice the energy of carbohydrates.

3. _____ is not a nutrient, but makes up most of your body weight.

4. _____ such as calcium and phosphorus are body-building nutrients.

5. _____ are the energy nutrients that are divided into sugar and starches.

6. _____ are needed for proper growth and aid in body repair.

B. List the four food groups.

1. _____

2. _____

3. _____

4. _____

32. Stay healthy

Anthony Potter

Define good health. _____

_____

How did you define good health? Most people mention maintaining a healthy body. Keeping your body fit and healthy is a big part of good health. There is another side to good health. That other side is your emotional health. Good health means being in good shape physically and emotionally.

Your physical health

There are six basic requirements to keeping a healthy body.

1. Exercise. Regular exercise helps keep your body fit. It strengthens muscles. It helps your circulatory system.

2. Cleanliness. Keep yourself physically clean. Every day disease-causing germs accumulate on your skin. A daily shower or bath controls the growth of these germs. It also rids the body of dirt and odor.

3. Rest. Get enough rest. Your body cannot function properly without enough rest. Most adults need seven to eight hours of sleep a night. Young children need even more. Relaxation is also important. Give your body a chance to relax between chores.

4. Nutrition. To stay in top form, your body must have the proper nutrition. A balanced diet will give you the nutrients you need. Nutrition was discussed in chapter 31.

5. Medical and dental care. You should see both your doctor and dentist regularly. Routine visits are important. Your doctor can spot diseases before they spread. Your doctor can give your shots to help prevent disease. Ask your doctor and dentist how often you should have exams.

6. Emotional health. Your emotional health is important to your physical health. If you are always tense, for example, you can develop health problems like ulcers or headaches or heart disease.

Spending time with friends and family can help you cope with stress.

Your emotional health

Your physical and emotional health affect each other. If you are in pain, for example, you are not likely to be in the best of moods! If you are feeling sad and depressed, you may feel tired and have little energy. You may develop aches and pains.

If your body feels healthy, you are better able to handle problems. And if you are taking care of your emotional health, you will have good energy. There are three basic requirements for maintaining good emotional health.

1. Learn to handle stress. **Stress** is tension or pressure. **Stressors** are the things that cause stress. Stressors can be happy or sad events. You cannot avoid all stressors. In fact, some stress is good for you.

Exercise and proper rest help your body cope with stress. Proper relaxation helps your mind handle stress. Many people have hobbies that help them to relax. Others meditate or take long walks or do exercise. Talking to friends and family about your feelings also helps you deal with stress. Don't let bad feelings pile up inside you. Letting bad feelings pile up can lead to emotional and physical problems.

2. Develop a positive self-image. What do you think of yourself? Do you like yourself? Everyone has good qualities. Be proud of yourself for who you are. Feeling good about yourself gives you more energy and enthusiasm for life. And the more energy and enthusiasm you have, the better you will keep feeling about yourself.

3. Take care of your physical health.

Chapter checkup

Name seven requirements for good health.

1. _____

2. _____

3. _____

4. _____

5. _____

6. _____

7. _____

Unit review

- A disease is an illness of the body.
- Most contagious diseases are caused by germs (viruses and bacteria).
- Most non-contagious diseases are caused by stress, drugs, poor diet, pollution, or birth defects.
- Antibodies are chemicals your body makes to fight foreign substances that enter your body.
- Tobacco, drugs, and alcohol can cause serious physical and emotional diseases.
- Addiction is dependence on a substance or behavior.
- Nutrients promote growth and supply energy.
- There are six nutrient groups: water, minerals, carbohydrates, fats, proteins, and vitamins. You need all of them for good health.
- There are four food groups: milk products, fruits and vegetables, bread and cereal products, meat (including nuts and beans).
- Exercise, eating right, and relaxing are important to good physical and emotional health.

Unit checkup

A. Tell the meaning of each word

1. epidemic _____

2. vaccine _____

3. addiction _____

4. nutritionist _____

5. stress _____

B. Fill in the answers

1. Name four diseases that are contagious.

_____

_____

_____

_____

2. Name four diseases that are not contagious.

_____

_____

_____

_____

3. Name the six nutrient groups.

_____

_____

_____

_____

_____

_____

C. Explain each statement

1. Physical health and emotional are related.

_____

_____

_____

2. A person's drug or alcohol addiction can affect their whole family.

_____

_____

_____

150

Unit 7.
The small earth

No plant or animal exists alone. Plants and animals rely on each other. Bees get nectar from flowers. While collecting the nectar, the bees help pollinate the flowers. There is balance to the world of nature.

Mankind's pollution and overuse of natural resources can upset the balance. Scientists are trying to find ways to clean up our damaged environment. Our well-being depends on how well we manage our resources. The science of ecology is helping us to learn how to take care of our natural resources and keep our environment healthy.

Larry Dacker USDA

Robert Maust

Pace USDA

Don Baldwin USDA

ecology
environment
ecologist
population
community
ecosystem
biome
biosphere

33. Ecology

Underwater environment

The word **ecology** comes from the Greek language. *Ecos* means "home." *Ology* means "the study of." Together the words mean "the study of homes." Ecology is the study of living things and their **environment**, their home.

A tiny drop of water may be the environment of an ameba. A cattail's environment is the pond where it is growing. A bluebird's environment is the woods it nests in.

Each organism must share its environment with many other living things. Look at the picture on page 153.

The bobcat must share its environment with the cows, the trees, and the fish. Think of a real pond. Think of all the different organisms that live there. They all live in the same environment. In doing so, they all must interact with each other. Ecology is the study of relationships between living things and their environments and with each other. An **ecologist** is a person who studies these relationships.

Populations, communities, and ecosystems

Ecologists sometimes choose to study one particular type of organism in an area. A **population** is all the members of one type of organism in an area. Ecologists might choose to study the population of oak trees in a forest. All the trout in a stream would be another population.

In any environment there are many different populations. In a pond there could be populations of frogs, waterbugs, and lily pads. These populations all interact with one another. The frogs sit on the lily pads and eat the waterbugs. All the populations in a particular area form a **community**.

Communities not only interact with each other, they also interact with their environment. Ecologists call an area where living organisms interact with each other and with their environment an **ecosystem**. An ecosystem may be as small as a drop of water. It might be as large as an ocean.

Giraffes are one of the many animal populations that thrive on the African grasslands.

Things to do

Use the drawings to answer these questions.

1. Name the different populations and give the number of organisms in each. _____

_____

_____

2. List the parts of this ecosystem that are non-living. _____

_____

_____

153

Evergreen forest

Biomes

An ecosystem is usually thought of as part of a larger region called a **biome**. A biome is a region in which a certain kind of plantlife lives. What can live in that biome depends on the amount of rain and sunlight, the kind of soil, competition from other organisms, and so on. Some of the most familiar biomes are deserts, grasslands, hardwood forests, evergreen forests, and tundras. Tundras are treeless plains in the Arctic region.

Most plants and animals have adapted to live in a particular biome. A desert cactus does not grow in the tundra. A polar bear lives in the tundra, not in the forest. But the same biome produces different species depending on where it is. The plants and animals that live in the desert of the southwest United States are different from those of the deserts of Africa.

There are a few living things that have adapted to life in just about any biome. They include humans, houseflies, and cockroaches!

The biomes are found on earth's **biosphere**. The biosphere is the thin outer crust of the earth and the lower layers of the atmosphere. Life on earth exists only in the biosphere.

Life on Earth exists in the biosphere.

Chapter checkup

Fill in the blanks with the following words:

population	community	ecology
ecosystem	environment	biome
biosphere		

1. _____ is the study of relationships among living things and their environment.

2. A desert, a forest, or a tundra is called a _____ by ecologists.

3. All the living organisms in a particular area form a _____.

4. The place where an organism lives is its _____.

5. A _____ is composed of all the members of one type of organism in a certain area.

6. An area where living and non-living things interact with each other is an _____.

7. Life exists on the thin outer crust of the earth called the _____.

34. Population

Population is all the members of one type of organism in an area. A population could be all the cats in New York City. All the cows in Farmer Brown's field are a population. All the roses in England are a population.

Things to do
Fill in the blanks to form three different populations:

1. All the _____ in _____.

2. All the _____ in _____.

3. All the _____ in _____.

Carl Purcell

Counting populations
Counting populations is often difficult. Counting animal populations is especially hard because animals move around. Some animals are counted twice and some are not counted at all.

Counting plant populations is usually easy. Let's say you wanted to know the maple tree population in a certain field. All you would have to do is count the trees. But what if you wanted to know the maple tree population in the state of Vermont? You couldn't count every maple tree in Vermont. *Try this #42* helps solve this problem.

Try this #42

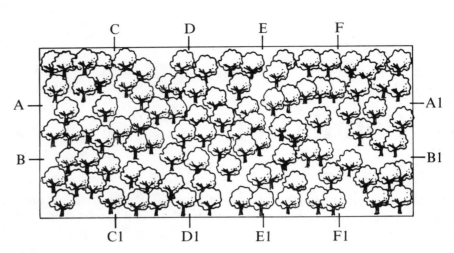

Materials: pencil, ruler

1. Estimate the number of trees in the field above (do not count.) Write your answer here. _____

2. Use your pencil and ruler to draw a line from A to A1. Do the same from B to B1, C to C1, D to D1, E to E1, and F to F1.

3. You have drawn 15 boxes. Count the number of trees in the upper lefthand box. Write your answer here. _____

4. Use the information from question 3 to estimate the number of trees in the field. Write your answer here. _____

You did not count all the trees. You counted the trees in a sample. You multiplied the number of trees in one section by the number of sections. From the sample, you determined the total number of trees. The answer probably will not be exact. But it will be a good estimate. This is the way ecologists determine the population in a large area.

Population controls

Nature has many ways to control population growth. The four most important population controls are...

Competition. Often several organisms will have the same needs. In a forest there are thousands of plants. They all need food, water, living space and sunlight. They are in competition with one another for these things. The strong will survive. The weak will die.

Predators. A **predator** is an animal that kills and eats other animals. The animals that are eaten by the predator are the **prey**. A wolf is chasing a rabbit. The wolf is the predator. The rabbit is the prey. Birds are very helpful predators. Many kinds of birds help control insect populations because insects are their prey.

Parasites. A **parasite** is a living organism that feeds on another living organism. There is a fungus that lives on elm trees. This fungus is a parasite. It grows on the elm tree. The fungus, in time, will kill the tree.

Disease. Diseases often infect living organisms. Sometimes the disease kills the organism. Sometimes the disease weakens the organism. The organism then becomes easy prey for predators or parasites.

The hawk is a predator of mice.

Things to do

Materials: 20 playing cards, 5 cardboard disks

You can learn about animal survival by playing this game. From your teacher get 20 playing cards and 5 cardboard disks. The playing cards stand for mice. Place five mice about 3 feet from your seat. The cardboard disks stand for hawks.

In order to survive, the hawks have to catch the mice. Toss the hawks at the mice, one at a time. Every time you hit a mouse, remove the mouse and leave the hawk on the floor. Every time you miss, add two mice to the floor and remove the hawk. Begin with 5 mice and 5 hawks. After each throw keep a total of the number of hawks and mice.

MICE									
HAWKS									

In the game, you acted as the hawk. When there were a lot of mice, they were easy to hit. The hawk population grew larger. The mouse population decreased. When there were fewer mice, they were harder to hit. Misses caused the hawk population to decrease. Misses caused the mouse population to grow.

Lack of food, bad weather, and pollution can affect animal populations.

In nature, a field full of mice makes hunting easy for hawks. The hawks thrive. Their numbers increase. In time, the number of hawks becomes too large. The hawks eat too many mice. The mouse population dwindles. There are not enough mice for all the hawks. Some of the hawks become weak and die. The hawk population begins to dwindle. As it does, the mouse population increases. There are more mice for the hawks. Eventually, the population of hawks grows again. The two species keep each other in balance.

Other population controls

The weather may kill large numbers of organisms. Drought (lack of rain), floods, or unusually cold or hot weather may kill plants and animals.

Humans also affect populations of plants and animals. Sometimes humans control populations on purpose. They cut down *some* trees in a forest so that other trees and plants and animals remain healthy.

But sometimes, humans affect populations in very negative ways, through carelessness or greed. Populations of plants and animals may be destroyed through fire or oil spills. People are sometimes predators. Humans hunted the buffalo to the point that the species almost disappeared from the earth. Other species have been wiped out. Pollution has killed many plants and animals.

Chapter checkup

A. Write **T** for true and **F** for false.

_____ 1. The only way to determine the population of catfish in the Mississippi River is to count them.

_____ 2. If a cat were chasing a mouse, the mouse would be the prey.

_____ 3. A population is all the different organisms in a certain area.

_____ 4. Diseases often weaken animals and make them easy prey.

_____ 5. In the hawks and mice game, the number of hawks dwindled because of pollution.

B. List four ways nature controls population.

1. _____

2. _____

3. _____

4. _____

35. Balance of nature

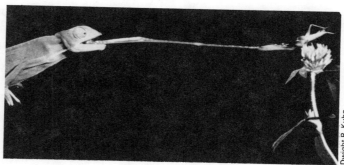

In chapter 34, you read about population controls. The controls of competition, predators, parasites, and disease are all part of the balance of nature. It is a system of checks and balances. The various organisms keep each other from overpopulating. They also provide each other with energy and needed nutrients.

Even the smallest organisms have a role in keeping the earth in balance. Bacteria keep the soil rich and fertile. Other small organisms are food to larger organisms.

Food chains

Organisms eat other organisms for energy. Organisms are part of **food chains**. There are many different food chains in each community. A grasshopper eats a plant. Then a toad comes along and eats the grasshopper. Soon, a snake comes along and eats the toad. Finally, a hawk swoops down and eats the snake. That is a simple food chain.

Food chains in a given area are connected. Most creatures eat a variety of organisms. And many plants and animals are food to many animal species. These various food chains form a food web.

Food Chain

A few of nature's cycles

The balance of nature depends on the environment's being in balance. The environment includes the water, the air, and the land, as well as the organisms. Animals and plants get nutrients they need from the environment. Plants get the water and nutrients they need from the earth. Animals get what they need from eating plants or other animals. So, if the balance is upset, the plants and animals in a food web are affected too.

These are some of the *cycles* that are part of the balance of nature.

Oxygen cycle. In unit 3, you read that plants make most of the oxygen that animals breathe. You also read that animals make carbon dioxide as a waste product. Plants need carbon dioxide in order to make food. Animals breathe the oxygen that plants make. Animals also eat the plants for food and for energy. Animals get much-needed nutrients from the plants.

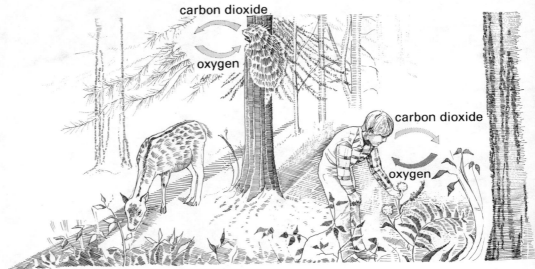

Oxygen Cycle

Nitrogen cycle. All living things need nitrogen. There is a huge supply of nitrogen in the air. Most plants can't use the nitrogen directly from the air. That nitrogen gas must be converted into nitrates in the soil. Here's how the cycle works:

1. The nitrate gets into the soil either through rain or through plant and animal decay and waste.
2. Plants take in nitrates dissolved in the water in the soil. These nitrates contain nitrogen.
3. Animals get nitrogen from eating plants or other animals.
4. Animals return some nitrogen through their wastes. They also return nitrogen to the soil when they die.
5. The animal or waste decays, and bacteria break down the nitrogen. It returns to the soil. Plants also decay and return nitrogen to the soil.
6. Plants take nitrates from the water in the soil.

Nitrogen Cycle

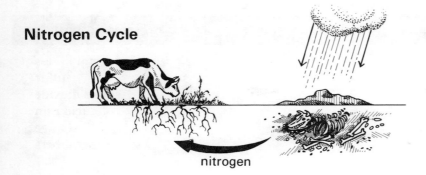

nitrogen

A water-cleaning cycle. Lakes, rivers, and oceans are kept clean by a water-cleaning cycle. When it is in balance, the cycle works like this:

1. Fish give off wastes.
2. Decay bacteria in the water break down the wastes into nitrates, phosphates, and carbon dioxide. They use up oxygen in the water when they do this.
3. Algae (tiny plants in the water) use sunlight to change these nutrients into food. The algae live off oxygen when they make the food. They provide oxygen for the water animals.
4. Small water animals eat the algae.
5. Fish eat the small water animals.
6. Fish give off wastes and the cycle repeats.

Food chains and cycles all work together to maintain a balance of nature, a balance of the natural world of plant, animal, and environment.

Where do people fit into this balance? We have come a long way from living in harmony with nature. Many people believe humans are in control of nature, rather than simply a part of it. Humans have changed the balance of nature. Sometimes, humans have destroyed that balance through pollution and overuse or abuse of the earth's resources. The balance has been upset by the use of pesticides. Killing one pest can lead to overpopulation of other plants and animals. We will look at pollution and the use of natural resources in the next two chapters.

Chapter checkup

Number the following steps to make a food chain.

a. _____ A plant makes its own food using the sun's energy.

b. _____ A fish eats the insect.

c. _____ A larger fish eats the fish.

d. _____ An insect eats the plant.

Number the following steps to show part of the nitrogen cycle.

a. _____ Bacteria cause decay of animal waste, dead animals, and plants.

b. _____ Plants absorb nitrates from the water in the soil.

c. _____ Animals leave waste materials, and other plants and animals die.

d. _____ Animals eat plants.

environmental illness
sulfur
sulfur dioxide
acid rain
sewage
scrubbers

36. Pollution

Pollution has made much of our air, land, and water dirty or unsafe. Pollution damages the environment. It has upset the balance of nature in some places. It also causes health problems.

Most pollution is caused by people. Our modern society uses a lot of chemicals and synthetic materials. Making these materials causes pollution of our land, air, and water. Sometimes, simply using these materials or being in the same room with them can make people sick. They have **environmental illnesses.** Environmental illnesses are health problems caused by chemicals in the air, water, or food. These illnesses even can be caused by building materials or household cleaning products. Symptoms of environmental illness can be headaches, depression, tiredness, stomach problems, and so on. Sometimes, it takes years for people to realize that something in the environment is making them sick.

Chemicals in our environment may be harmful to our health.

Upsetting the balance

In nature, the waste products of a plant or animal are used by other plants or animals. Everything is recycled. The cycles of nature go on.

But waste is a problem in the human world. We have taken minerals from the earth and changed them into new forms. We are making materials, like plastics and many chemicals, that do not break down and become useful to nature again. Some harmful materials collect in our environment.

Pollutants combine with fog to form smog over Hollywood, California.

Air pollution

Burning of fuel is the major source of air pollution. We burn fuel to heat our homes, to manufacture products, and to run our cars. Much of our air pollution comes from cars, trucks, and buses.

Cars burn gasoline. They give off exhaust. Exhaust contains many poisonous chemicals that contaminate the air. Exhaust from cars can cause headaches, nausea, and stomach cramps.

Most homes are heated by burning oil or natural gas. Many factories burn oil, natural gas, or coal for their energy needs. These fuels contain a chemical called **sulfur**. This sulfur combines with oxygen to create **sulfur dioxide**. Sulfur dioxide irritates people's skin and eyes. It causes breathing difficulties. When it rains, the sulfur dioxide combines with the rain water. This is called **acid rain**. Acid rain is very harmful to animals and plants. In the northeastern United States and Canada, acid rain has killed the plant and animal life in many lakes and streams.

Sometimes pollutants in the air combine with fog. This is called smog. Smog is a dangerous health hazard. In 1952, a very thick smog hung over London, England. It resulted in the deaths of more than 4,000 people.

Cigarette smoking is also a form of air pollution. Air that has cigarette smoke in it is called secondhand smoke. Doctors now say that secondhand smoke can cause cancer and breathing problems for nonsmokers.

"GEE BOSS, THANKS FOR THE LIFT HOME"

Sometimes in hot weather, city people with heart and breathing problems must stay indoors; the air is not safe for them. Warm air moves over the polluted air of the city. The pollution can't rise through the warm air. The warm air acts as a lid. It holds the polluted air down over the city. This is called temperature inversion.

Air pollution is not limited to large cities. It affects small towns and farms. It affects wilderness areas where there are no cars and factories. Smog has been seen over the oceans and over the North Pole. Scientists say a thin layer of pollution hangs over the whole earth.

Scientists are predicting some frightening long-term effects from the burning of too much fuel. They say that too much carbon dioxide is being released into the air. Some scientists predict a "greenhouse effect" that will cause a warming of the earth. Carbon dioxide absorbs infrared heat from the sun. The earth will warm because the infrared heat won't be able to escape the atmosphere. This could lead to melting of the ice caps at the North and South Poles. This would raise the level of the oceans and cause flooding. The entire earth would be affected as the climate changed.

Other scientists, however, say that there will be a cooling of the earth from all the pollution. They say the pollution will reflect the suns rays away from the earth's atmosphere. This cooling would also affect the earth's climate. It could cause another Ice Age, scientists say. Either way, there could be huge changes in the environment.

Car exhaust is a major cause of air pollution.

Try this #43

Materials: 5 white index cards, a jar of petroleum jelly, tape, magnifying glass

1. Smear a thin layer of petroleum jelly on the center of each card.
2. Tape the cards to a number of different locations, such as on a tree, on a wall, or on a window sill.
3. Label each card with its location.
4. Leave the cards.
5. After two or three days, collect the cards. Be careful not to disturb the petroleum jelly.
6. Use your magnifying glass to examine any particles stuck in the petroleum jelly.

When you looked at your cards, were you surprised at the number of particles that stuck to the petroleum jelly? Much of what you see on your cards is air pollution. Some may be natural pollutants, such as pollen. But much of the dirt and dust is caused by cars and factories. Compare your cards with other people's cards. Why are some cards dirtier than others? Is there a difference in the types of particles collected?

Chemicals in polluted water kill fish and water plants.

Water pollution

Clean water is important to everyone. We can't live without it. Water pollution is caused by waste materials in the water that make the water harmful to living organisms. People are the cause of most water pollution.

Human waste is called **sewage**. In some places sewage is dumped directly into rivers or lakes. Bacteria live in sewage. They make the water unsafe for drinking because they can cause disease. They also need oxygen in order to live. They use up a lot of the oxygen in the water. Plants and animals in the water die because they can't get enough oxygen.

Chemicals also pollute water. Every day we use chemicals in our homes. Farmers use chemicals as fertilizers. Factories use chemicals to help make their products. Many of these chemicals end up in our lakes and streams. They kill plant and animal life.

164

Things to do

Mercury is a chemical that pollutes water. Bacteria that live in the water absorb mercury with their food. In the space below, explain how this mercury could become a human health hazard. (Hint: Many protists use bacteria for food.)

_____

_____

Heat is a form of water pollution. Factories often use water to cool their machines. The machines take in cool water and return warm water. Warm water dumped directly into streams and lakes reduces the oxygen supply. Animals that need a lot of oxygen may die. Heat pollution disrupts the life cycles of many fish. It can cause fish eggs to hatch too soon or not hatch at all.

In recent years, there have been many oil spills into the world's oceans. The oil is spilled from the ships and rigs of the oil companies. This oil has killed fish, birds, mammals, and small organisms. It also reaches beaches around the world. Sewage and industrial waste also have been dumped into the ocean.

Land pollution

Trash is the major cause of most land pollution. More than 500,000 tons of trash are produced in the United States every year. The disposal of this trash is a major problem. Some places burn their trash. But this puts pollutants into the air because much of the trash is made of plastics and other synthetic materials. Some coastal cities dump their trash into the sea. This pollutes the water. Now most trash is being buried in landfills. But this also creates problems. Landfills generally have a bad odor. They can pollute the air. Sometimes harmful chemicals are buried. These chemicals may leak into the local water supply. The water supply is poisoned. The poisoned water can cause cancer, other illnesses, and birth defects.

Fertilizers and insecticides (chemicals to kill insects) pollute the soil. These chemicals get into the water supply. They also get into the foods we eat. DDT was one of the insecticides farmers once used. It is now outlawed. But this and other chemicals can still be a danger to people and animals. Animals store DDT in their cells. DDT can affect the plants and animals in a food chain. Suppose an animal eats a treated plant. The DDT gets into the animal's system. A predator comes along and eats the first animal. Now the predator has DDT in its system. And so the chemical travels up the chain.

Radiation

Some radiation is produced by nature. But humans are creating radiation through x-rays, nuclear power, and nuclear weapons. Officials say there is little risk of radiation being released into the air or water supply. Other people disagree. In 1986, there was a nuclear accident in the Soviet Union. Radiation was carried thousands of miles. Animals and plants were exposed to the radiation. Scientists are still not sure how far-reaching the effects will be.

Finding ways to get rid of garbage is a serious problem for many cities.

165

Toxic wastes being stored

Jeffrey High

The use of nuclear power has created another problem—what to do with nuclear waste. There is no safe way to store it or get rid of it. Some nuclear waste remains dangerous for thousands of years. Nuclear waste is carried in trucks on our highways to storage areas. Many people worry about what will happen if these trucks are in an accident.

Solving the problems

Solving the pollution problem is not easy. Efforts are being made to produce cars with less harmful exhaust. Many factory smoke stacks are now being equipped with **scrubbers**. Scrubbers remove the sulfur dioxide from the smoke.

In many places in the United States, it is against the law to dump any polluting substance into rivers or lakes. Most waste water is now being cleaned at water treatment plants.

Landfills are being monitored to help prevent harmful chemicals from being buried. Some cities are burning their trash. These new trash-burning plants are equipped with scrubbers. They use the heat energy to produce electricity. Many places are recycling their trash.

The pollution problem is far from being solved. Acid rain still plagues the Northeast. This problem spans two countries as it is also destroying much of Canada's waterlife. Despite laws against it, many factories keep dumping hazardous chemicals into lakes and streams.

Pollution problems are complex and dangerous. Citizens, private industry, and the government are trying to find answers they can afford. They are balancing the costs against the benefits of a clean and safe environment.

Chapter checkup

List six effects of pollution on people or the environment. Write something more specific than simply "air pollution," or "water pollution."

1. _____

2. _____

3. _____

4. _____

5. _____

6. _____

37. Natural resources

Soil is removed to mine coal. Coal is an important natural resource.

The earth gives us many gifts. We call these gifts **natural resources**. Natural resources are the natural substances that we use. Air, water, and minerals are natural resources. Plants and animals are natural resources. Coal and oil are natural resources.

Renewable and non-renewable resources

Many natural resources are **renewable**. Renewable resources can be replaced in a short period of time. Plants and animals are renewable resources. They replace themselves by reproducing. Air and water are renewable resources. They are constantly being recycled.

There are many natural resources that are not renewable. The earth's minerals are non-renewable. There is only a limited supply. Once they are used up they are gone forever. Coal and oil are considered non-renewable resources. Topsoil is considered a non-renewable resource.

Things to do

Coal, oil, and topsoil can actually be replaced by the earth. Why do you think the ecologists call them non-renewable?

Strip mining

Using our natural resources

There is only a limited amount of land on the earth. People need this land for living and for growing food. As the population increases, so does the amount of land we use. As humans inhabit more and more land, there is less and less room for other animals and plants. Animals are driven from their homes. Plants are uprooted. Many animal and plant species are in danger of disappearing because of the actions of humans. People are now asking environmental questions before starting large building projects. "Are we upsetting the ecology of our area by building? What are we destroying? Is it worth it?" It is hard to know just how much our actions today will affect the earth 100 or 1,000 years from now.

There are efforts now to preserve wildlife. There are fish and game laws. These laws limit the number of animals killed by fishers or hunters. There are wildlife preserves where no fishing or hunting can take place. There is land set aside that is to remain "forever wild." Even that land, however, is affected by air and water pollution.

When we say air is a natural resource, we mean clean air. Polluted air is distasteful to breathe. It is unhealthy. Clean water is also a natural resource. Are clean air and water becoming non-renewable natural resources? We must work to keep our air and water pure.

Plants and trees supply us with oxygen, food and may other things. Plants and trees also hold topsoil in place. When they are removed **soil erosion** takes place. Soil erosion occurs when wind or water sweeps away the topsoil.

Strip mining has destroyed thousands of acres of good land. Trees, plants, and soil have been bulldozed in order to get at the minerals underneath. The land has been left bare. Topsoil has been washed away. Most soil erosion can be prevented with careful planning.

Trees will be planted to replace the trees removed from this forest.

Materials: pan, topsoil, sod, watering can
1. Place about two handfuls of topsoil in the pan. Move the topsoil to one end to form a small mound.
2. Next to the topsoil place a clump of sod.
3. Tip the pan up slightly.

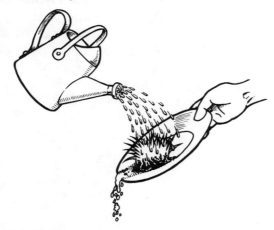

4. Use the watering can to pour water on the elevated end so that half falls on the grass and half falls on the topsoil.
5. What did you observe happening to the topsoil?

_____

6. Did the same thing happen on the grassy side?

_____

Water washed away the loose topsoil. Topsoil held in place by the grass did not wash away. Farmers must be very careful when they plow a field so their topsoil is not washed away. Contour plowing and terracing are ways farmers help preserve the topsoil.

Fertile soil has been destroyed in other ways. Farmers have planted the same crop over and over again on their land. This has stripped the soil of nutrients. Today, many farmers rotate their crops. They plant different crops from year to year in order to renew the soil.

Farmers have also been using chemical fertilizers instead of animal wastes (manure) on their fields. The animal wastes put nutrients back into the soil. Then chemical fertilizers do not renew the soil. The fertilizers can be washed away. The farmer must use more and more chemicals to keep crop production up. The soil becomes depleted (used up). Chemicals are also used to kill insects and other pests. The chemical fertilizers and pesticides (pest killers) also seep into the ground water and pollute it. They also get absorbed into the foods the farmers grow and the animals they raise. Then people eat these chemicals when they eat the plants and animals.

This aircraft is spraying pesticides on a corn crop. The pesticide helps the crop to grow, but can harm animals and plants.

169

Newspaper is turned into pulp and then made into new paper at this recycling plant.

Conserving our natural resources

Iron, copper, and gold are three of the earth's many different minerals. Minerals are presently being mined at a very rapid rate. Someday they will all be gone. We must learn to conserve these minerals. One way is recycling. To **recycle** means to treat or process something so that it can be used again. Most minerals can be recycled.

Coal and oil are non-renewable resources. Most of the energy we use today comes from coal or oil. We must learn to conserve energy. One way to conserve energy is by driving less often or carpooling. We can conserve energy by using less electricity. We can look for energy from other sources. Other forms of energy include solar, wind, and water power. Some people believe we should turn to nuclear power for our energy.

Conserving our natural resources is everyone's job. If we all do our part, our natural resources will last a long time.

Chapter checkup

A. List eight natural resources. Next to each, state whether it is renewable or nonrenewable.

1. _____

2. _____

3. _____

4. _____

5. _____

6. _____

7. _____

8. _____

B. List five things that can be done to help preserve natural resources.

1. _____

2. _____

3. _____

4. _____

5. _____

Unit review

- Ecology is the study of the relationship of living things and their environment.
- All the populations in a given area are a community.
- A biome is a region in which a certain kind of plantlife lives. Deserts, tundras, forests, and grasslands are biomes.
- Life exists only in the biosphere.
- Populations can be counted either directly or by taking samples.
- Population is controlled by competition, predators, parasites, and disease.
- Living things are parts of food chains.
- There are nature cycles that recycle water, oxygen, and other nutrients.
- Air, water, and land pollution can upset nature's balance.
- Most pollution is caused by burning fuel for driving, heating our homes, and making products.
- Disposing of trash and industrial waste products is a critical environmental problem.
- There are renewable and non-renewable natural resources.
- Soil erosion is a conservation problem.

Unit checkup

Fill in the blanks

environment   pollution
ecosystem   natural resources
biome   community
biosphere   ecology

1. An organism's home is its _____.

2. A _____ is a region with a certain plantlife.

3. Life exists only in the earth's _____.

4. The study of living things and their environment is _____.

5. A fish tank or a terrarium is an example of a small _____.

6. Coal, plants, topsoil, and water are _____.

7. All the populations in a given area make up a _____.

8. _____ can make an environment unsafe and unclean.

True or false

_____ 1. Burning of fuel is the major cause of air pollution.

_____ 2. Chemical fertilizers and pest killers get into the food chain.

_____ 3. Plastics and other synthetic materials break down to become part of nature again.

_____ 4. Nature recycles water and oxygen.

_____ 5. Cutting down forests is one way to conserve topsoil.

_____ 6. Chemical wastes can safely be buried in the ground.

_____ 7. Oxygen and nitrogen cycles help keep nature in balance.

_____ 8. The loss of a species of plant or animal could affect the balance of nature years from now.

_____ 9. Sulfur dioxide combines with rain to form acid rain.

_____ 10. Acid rain helps keep topsoil rich and fertile.

_____ 11. Acid rain helps keep topsoil rich and fertile.

Joseph H. Starowicz, an English teacher in the Syracuse City School District, has worked in and helped develop various remedial programs for underachievers. He has a B.A. in English from Le Moyne College and an M.S. in Education from Syracuse University. Throughout his 12 years of teaching, Starowicz has been involved with various committees working to enhance the quality of public education. He is a member of the American Federation of Teachers and the National Science Teachers Association.

Stephen A. Martin B.S, M.S., is a graduate of Florida Institute of Technology with a major in mathematics. He is certified to teach both secondary mathematics and science. In addition to teaching in the Syracuse area schools, he has served on the committee to identify and aid the handicapped learner.

Martin authored "A Tale of Two Shepherds," published by *Curriculum Review*, which is a hypothetical account of the development of our present-day number system.